Mayne Reid

Bruin

The Grand Bear Hunt

Mayne Reid

Bruin
The Grand Bear Hunt

ISBN/EAN: 9783337217631

Printed in Europe, USA, Canada, Australia, Japan

Cover: Foto ©berggeist007 / pixelio.de

More available books at **www.hansebooks.com**

THE GRAND BEAR HUNT.

BY

CAPTAIN MAYNE REID,

AUTHOR OF "THE BOY HUNTERS," "THE YOUNG VOYAGEURS,"
"ODD PEOPLE," ETC., ETC.

BOSTON:
TICKNOR AND FIELDS.
M DCCC LXI.

AUTHOR'S NOTE.

CAPTAIN REID acknowledges with pleasure the assistance of an American Author, the results of whose labors he has been kindly enabled to incorporate with his own in the story of "Bruin: or, The Grand Bear Hunt."

University Press, Cambridge :
Stereotyped and Printed by Welch, Bigelow, & Co.

CONTENTS.

iv CONTENTS.

CONTENTS.

LIV.	A SLEDGE-CHASE	302
LV.	THE SUN BEARS	308
LVI.	THE TALL TAPANG	313
LVII.	THE BRUANG	318
LVIII.	THE CABBAGE-EATER	324
LIX.	THE SLOTH BEAR	332
LX.	BRUIN TAKEN BY THE TONGUE	339
LXI.	AN EXTRA SKIN	345
LXII.	AN UNHAPPY HORSE	351
LXIII.	THE SNOW BEAR	356
LXIV.	THE LAST CHASE	363

BRUIN.

CHAPTER I.

THE PALACE GRODONOFF.

On the banks of the Neva, near the great city of St. Petersburg, stands a splendid palace, known as the Palace Grodonoff. It is the property of a Russian nobleman of that name, as it is also his place of residence. Were you to drive up to the front gate of this grand palace, you would see a coat-of-arms sculptured in granite over the entrance. In this piece of sculpture, the principal and most striking figure is a bear, with the blade of a knife buried in his breast, the haft being clutched by a human hand! Open the gate, and enter the spacious court-yard. Inside, on the right and left, you will observe two live bears — both of chestnut-brown color, and each of them as big as a buffalo. You cannot fail to notice them, for, ten chances to one, they will rush towards you with fierce growls; and were it not that a strong chain hinders them from reaching you, you might have reason to repent having entered the court-yard of the palace Grodonoff. Look around you in the court-yard and over the different doors that open upon it; you will again see the crest of the bear,

1   A

sculptured in **stone;** you will see **it over the** stables, the
coach-house, the granary, the kitchens, — everywhere.
You may know by all this, that it is the coat-of-arms of **the**
Baron Grodonoff, whose crest is a bear with a blade **bur-**
ied in its breast, and **a** human hand clutching the haft.

You will naturally conclude that there is some history
connected with this singular **tableau** — that it is the com-
memoration **of some** deed done by a **Grodonoff,** entitling
him to use the bear as **his** heraldic device. **This** is quite
true; and **if** you enter the picture-gallery of the palace,
you will **there** behold **the deed more** explicitly represented, in a large oil-painting hung conspicuously in the
centre of the wall. The scene of this painting is a forest of old trees, whose gray, gnarled **trunks stand** thickly
over the ground. There is only a little open space or
glade in the middle; and this **is** occupied by three
figures, two men and a **bear.** The bear is between the
two men; or, rather, one of the men is prostrate upon
the ground — where he has been struck down by a blow
from Bruin's paw — while **the** huge animal stands over
him reared up on his hind-quarters. The other man is
upon his feet, apparently engaged in a desperate wrestle
with the fierce brute, and likely to prove the conqueror
— as he has already buried the blade of a large hunting-
knife in the animal's breast, and directly over the region
of its heart. Indeed, the shaggy monster **already shows**
signs of succumbing. His paw has dropped from the
shoulder of his antagonist, his long tongue lolls out, the
blood rushes from his mouth and nostrils, and it is evident that his strength is fast forsaking him, **and** that he
will soon sink lifeless upon the earth.

You will notice that the two men who figure in the painting are very dissimilar in appearance. Both are young men, and both are in hunting-costume; but so unlike is their dress, that you could not fancy they followed the same occupation. He upon the ground is richly attired. He wears a tunic of finest green cloth slashed with sable-fur on the skirt, collar, and sleeves; his limbs are encased in breeches of white doeskin; and his boots, reaching nearly to his thighs, are of soft russet leather, ample at the tops. A belt around his waist is richly embroidered; and the hilt of a short hunting-sword, protruding from the sheath, appears chased and studded with jewels. A light plumed hat lies upon the ground near his head — evidently tossed off in the struggle — and beside it is a boar-spear that has been jerked out of his fingers as he fell. The whole costume is similar to that used upon the stage — when some young German or Sclavonian prince is represented as hunting the wild boar in the forests of Lithuania.

In reality it is a prince who is depicted in the group of the gallery Grodonoff — but not a German prince. He is a Russian, and the bear is the Russian bear.

The other hunter — he who has given its death-blow to the fierce quadruped — is dressed in a style entirely different. It is the costume of a fur-hunter — a trapper of sables — and consists of skin coat and cap, with a strong leathern belt round his waist, and rough boots of untanned hide upon his legs and feet. The costume is rude, and bespeaks him a peasant; but his face, as the painter has represented it, is neither common nor ill-looking. It is not so handsome as that of the prince:

for he would be an unskilful artist — one utterly reckless
of his own fortune — who should paint the features of a
peasant as handsome as those of a prince. In Russia,
as elsewhere, such an imprudent painter would be a *rara
avis* indeed.

The picture of which we are speaking is the *pièce de
resistance* of the Grodonoff gallery. Its size and conspicuous position declare the fact; and the story attached
to it will show that it merits the distinction. But for
that picture, or rather the scene which it represents,
there would be no Grodonoff gallery — no palace — no
baron of the name. Paintings, palace, title, all have
their origin in the incident there represented — the battle
with the bear.

The story is simple and may be briefly told. As
already stated, he upon his back, hat off, and spear de-
tached from his grasp, is a Russian prince — or rather
was one, for at the time when our history commences he
is an emperor. He had been hunting the wild boar;
and, as often happens to sporting princes, had become
separated from his courtier attendants. The enthusiasm
of the chase had led him on, into the fastnesses of the
forest, where he came suddenly face to face with a bear.
Princes have their hunter ambition as well as other men;
and, in hopes of taking a trophy, this one attacked the
bear with his boar-spear. But the thrust that might
have penetrated the flesh of a wild boar, had no effect
upon the tough, thick hide of Bruin. It only irritated
him; and, as the brown bear will often do, he sprang sav-
agely upon his assailant, and with his huge paw gave the
prince such a "pat" upon the shoulder, as not only sent

the spear shivering from his grasp, but stretched his royal highness at full length upon the grass.

Following up his advantage, the bear had bounded forward upon the prostrate body; and, no doubt, in the twinkling of a bedpost would have made a corpse of it — either squeezing the breath out of it by one of his formidable " hugs," or tearing it to pieces with his trenchant teeth. In another moment the hope of Russia would have been extinguished; but, just at this crisis, a third figure appeared upon the scene — in the person of a young hunter — a *real* one — who had already been in pursuit of the bear, and had tracked him up to the spot.

On coming upon the ground, the hunter fired his gun; but, seeing that the shot was insufficient, he drew his knife and rushed upon the bear. A desperate struggle ensued, in which, as may be already anticipated, the young hunter proved victorious — having succeeded in sheathing his blade in the heart of the bear, and causing the savage quadruped to " bite the dust."

Neither the prince nor the peasant came scathless out of the encounter. Both were well scratched; but neither had received any wound of a serious nature; and the amateur hunter rose once more to his feet, conscious of having made a very narrow escape.

I need not add that the prince was profuse in his expressions of gratitude to him who had saved his life. The young hunter was not one of his own party, but a stranger to him, whose home was in the forest where the incident occurred. But their acquaintance did not end with the adventure. The prince became an emperor, — the peasant hunter a lieutenant in the Imperial Guard,

afterwards a captain, a colonel, a general, and finally a
baron of the empire!

His name?

Grodonoff, — he in whose palace hangs the picture
we have described.

CHAPTER II.

THE BARON GRODONOFF.

In one of the apartments of the palace Grodonoff, behold its proprietor, the baron himself! He is seated in an old oak chair, with a heavy table of the same material in front of him. On the table is spread out a map of the world; and by the side of the chair stands a large terrestrial globe. Several shelves standing against the walls contain books; and yet the apartment is not a library, in the proper sense of the word: rather is it a large oblong saloon; having three of its sides occupied by spacious glass cases, in which are exhibited objects of natural history, — birds, quadrupeds, reptiles, and insects, — all mounted in proper form, and arranged in due order. It is, in fact, a museum, — a private collection — made by the baron himself; and the books that fill the shelves are works relating to natural history alone.

In a man of military aspect, — an old veteran with snow-white hair, and grand moustaches of like hue — such as he who is seated at the table — you would scarce expect to meet the lover of a study so pacific in its character as that of natural history? Rather would you look to find him poring over plans of fortifications, with the pages of Vauban spread open before him; or

some history detailing the campaigns of Suwarrow, Die-
bitsch, Paskiewitch, or Potemkin? In this instance,
however, appearances were deceptive. Though the
baron had proved an excellent military officer, and seen
service, he was a student of **Nature.** His early life,
spent as a hunter, had begot within him a taste for natural history; which, as soon as the opportunity offered,
had become developed by study and research. It was
now no longer a predilection, but a passion; and in his
retirement nearly the whole of his time was devoted to his
favorite study. A vast fortune, which his grateful sover-
eign had bestowed upon him, enabled him to command
the means for gratifying his taste; and the magnificent
collection by which he was surrounded gave evidence
that no expense was spared in its gratification.

It was a map and globe of the world that now occupied his attention. Could these have reference to a
question of natural history?

In an indirect manner they had, — and what follows
will account for their presence.

A hand-bell stood upon the table. The baron rang
it; and before its tinkling had ceased, the door opened,
and a servant entered the apartment.

" Summon my sons to attend me!"

The servant bowed, and retired.

A few minutes after, two youths entered the apartment. They appeared to be of the respective ages of
sixteen and eighteen. One, the elder and taller, was of
a darkish complexion, with brown, waving hair, and
hazel eyes. The expression of his countenance was
that of a youth of firm and rather serious character;

while the style of his dress, or rather his manner of wearing it, showed that he was altogether without vanity in matters of personal appearance. He was handsome withal, having that aristocratic air common to the nobility of Russia. This was Alexis.

The younger brother differed from him as much as if no kinship existed between them. He was more the son of his mother, the baroness; while Alexis inherited the features and a good deal of the disposition of his father. Ivan was a fair-haired lad, with golden locks curling over a forehead of bright blonde complexion, and cheeks that exhibited the hue of the rose. His eyes were of a deep azure-blue — such as is often seen among the Sclavonic races — and their quick sparkle told that in the breast of Ivan there beat a heart brimming with bright thoughts, and ever ready for mischief and merriment, but without any admixture of malice.

Both approached their father with a serious expression of countenance. That of Alexis bespoke sincerity; while Ivan stole forward with the air of one who had been recently engaged in some sly mischief, and who was assuming a demure deportment with the design of concealing it.

A word about these two youths, and the object for which their father had summoned them into his presence. They had now been each of them more than ten years engaged in the study of books, under some of the ablest teachers that Russia could furnish. Their father himself had given much time to their instruction; and, of course, an inclination to their minds similar to that which characterized his own, but chiefly to the mind of Alexis.

1 *

The latter had imbibed a fondness for the study of nature, while Ivan was more given to admire the records of stirring events, with a strong *penchant* for the splendors of the world, in which he felt longing to bear a part. The nature of the books which had passed through their hands — a great number of them being books of travel — had begotten within these youths a wish to see the world, which, increasing each day, had grown into an eager desire. This desire had been often expressed in hints to their father; but at length, in a more formal manner, by means of a written petition, which the boys, after much deliberation, had drawn up and presented to him, and which was now seen lying open before him upon the table.

The petition was simply their united request that their father would be so good as to allow them to travel and see foreign countries, — where, and how, to be left to his wise guidance and discretion.

It was to receive an answer to this petition, that his sons were now summoned into his presence.

CHAPTER III.

THE SEALED ORDERS.

"So, my youngsters!" said the baron, directing his glance upon them, "you have a desire to travel? You wish to see the world, do you?"

"True, papa!" modestly answered Alexis; "our tutor tells us that we are sufficiently educated to go abroad; and, if you have no objection, we should very much like to make a tour."

"What! before going to the University?"

"Why, papa! I thought you were not going to send us to the University for some time to come? Did you not say that a year of travelling was worth ten at a university?"

"Perhaps I may have said so; but that depends upon *how* one travels. If you travel merely to amuse yourselves, you may go over all the world, and come back no wiser than when you started! I have known many a man return from a circumnavigation of the globe, without bringing with him the knowledge of a single fact that he might not have obtained at home. You would expect to travel in snug railway carriages, and comfortable steamships, and sleep in splendid hotels — is that your expectation?"

"O no, papa! whatever way you may direct, that will be agreeable to me," said Alexis.

"As for me," rejoined Ivan, "I'm not particular. I can rough it, I'm sure."

There was a little flavor of *bravado* in the manner of Ivan's speech, that showed he was scarce inclined to the roughing system, and that he merely assumed the swaggering air, because he had no belief that he would be called upon to make trial of it.

"If I permit you to travel," continued the baron, "where would you like to go? You, Alexis! to what part of the world would your inclination lead you?"

"I should like to see the new world of America — its noble rivers, and forests, and mountains. I should certainly visit America, if it were left to my choice; but I shall be guided by you, papa, and do as you direct."

"You, Ivan?"

"Paris for me, of all places in the world!" replied Ivan, without any suspicion that the answer would be displeasing to the father.

"I might have **known so," muttered the** baron, with a slight frown clouding his forehead.

"O papa!" added Ivan, noticing the shade of displeasure which his answer had produced; "I don't care particularly about Paris. I'll go anywhere — to America, if Alexis likes it best — *all round the world* for that matter."

"Ha! ha! ha!" laughed the baron; "that sounds better, Ivan; and, since you offer no objection to it, *all round the world you shall go.*"

"Indeed? I'm glad to hear it," said Alexis.

"What! visit all the great cities of the world?" exclaimed Ivan, whose mind was evidently occupied with the delights of great cities.

"No," replied his father; "it is just that which I do not intend you shall do. There is a great deal to be learnt in cities, but much that would be better not learnt at all. I have no objection to your passing through cities, — for you must needs do so on your journey — but one of the conditions which I shall prescribe is, that you make stay in no city longer than you can arrange for getting out of it. It is through *countries* I wish you to travel — amidst the scenes of nature — and not in towns and cities, where you would see very little more than you can in St. Petersburg itself. It is Nature I wish you to become acquainted with, and you must see it in its most primitive forms. There only can you appreciate Nature in all its sublimity and grandeur."

"Agreed, papa!" exclaimed both the boys at once. "Which way do you wish us to go?"

"All round the world, as Ivan has suggested."

"O, what a long voyage! I suppose we shall cross the Atlantic, and then by the Isthmus of Panama to the Pacific; or shall we go as Magellan went, around Cape Horn?"

"Neither way — I wish you to make great journeys by land, rather than voyages by sea. The former will be more instructive, though they may cost more time and toil. Remember, my sons! I do not send you forth to risk your lives without a purpose. I have more than one purpose. First, I wish you to complete your studies of natural science, of which I have taught you the elements.

The best school for this is the field of Nature herself which you shall explore in your travels. Secondly, as you both know, I am fond of all natural objects, but especially those that have life — the beasts of the field, and the birds of the air; these you must observe in their native haunts, with their habits and modes of existence. You will keep a journal of all facts and events that may be worth noting down, and write out in detail such adventures as may occur to you upon your route, and you think may prove interesting to me, to read on your return. I shall provide you with ample means to accomplish your journey; but no money is to be wasted by idly sojourning in large cities: it must be used only for the necessary expenses of your travels. The emperor has been kind enough to give you a circular letter, which will get you funds and such other assistance as you may require from his agents in all parts of the world.

"We promise, dear father, strictly to adhere to your instructions. But whither do you desire us to go?" Alexis asked the question.

The baron paused for some time before making reply. Then, drawing from his desk a sealed paper, which showed signs of having been but recently folded, he gravely said, as he held it towards them, —

"In this document you will find the conditions upon which I give you permission to travel. I do not ask you to agree to them until you have carefully examined and reflected upon them. You will therefore retire to your room, read this document over, and, having given its contents due consideration, return, and signify whether you accept the terms; for if not, there is to be no travelling."

"*By the Great Peter!*" whispered Ivan to Alexis, "they will be hard, indeed, if we don't accept them."

Alexis took the paper, and both, bowing to their father, retired to their own apartment.

The seal **was** immediately broken, and not without some surprise did they peruse the contents of the document. It was in the form of an epistle, and ran thus : —

"MY SONS ALEXIS AND IVAN:

"You have expressed a desire to travel, and have requested me to give you my permission. I accede to your request, but only upon the following conditions: You must procure for me a skin of every variety of bear known upon the earth. I do not mean such varieties as are termed 'accidental,' arising from albinism or like circumstances, but every species or variety known to naturalists and acknowledged as 'permanent.' The bears from which these skins are to be taken must be killed in their native haunts, **and** by your own hands, — with no other assistance than that of an attendant whom I shall appoint to accompany you. In order to accomplish the task which I have imposed upon you, it will be necessary for you to go 'round the world;' but I add the further condition, that you **are** to go *only once round it.* In *latitude*, I leave you free to range, — from pole to pole, if it so please you [this was a stretch of liberty at which both boys laughed]; but *longitudinally*, no. You must not cross the same meridian twice before returning to St. Petersburg. I do not intend this condition to apply to such traverses as you may be compelled to make, while actually engaged in the chase of a bear, or in tracking the animal to his den; only when you are *en route* upon your journey. You will take your departure from St. Petersburg, and go east or west, which you please. From the conditions I have imposed upon you, I hope **you** will have skill enough to discover that a route is traced out for you, and, that, on starting, you *can* follow it either eastward or westward. This, with all matters relating to your means and mode of travelling, I leave to your own choice; and I trust that the practical education you have received will enable you to make your tour with proper judgment. ['Tour, indeed!' exclaimed Ivan.] Once out of my palace, I take no further charge of you. You may be some years older before I see you again; but I trust the time will not be misspent; and that upon your return you may be able to give a proper account of yourselves is the earnest hope and wish of your affectionate father,

"MICHAEL GRODONOFF."

CHAPTER IV.

DISCUSSING THE CONDITIONS.

THE two youths were no little astonished by the contents of this singular epistle; but, for all that, the terms imposed did not seem to them either harsh or unreasonable, and they were only too pleased to accede to them. They partly guessed their father's motive. They knew that he loved both of them with a true paternal love; but his affection was not of that kind to pet and pamper them within the precincts of his luxurious palace. He had a different idea of what would be beneficial to their future interests. He believed in the education which is acquired in the rude school of toil and travel, more than in the book-lore of classic universities; and he was determined that they should have a full measure of this sort of training. He had resolved that they *should* see the world, — not according to the ordinary understanding of this hackneyed phrase — not the world of towns and great cities, with their empty shows and vices — but the *world of Nature;* and, in order that they should have the opportunity of becoming thoroughly acquainted with this sort of world, he had traced out for them a route that would lead them into its very wildest scenes, and disclose to them its rarest and most primitive forms.

B

"By my word, brother!" exclaimed Ivan, as soon as Alexis had finished reading the letter, "we shall have travelling to our hearts' content. Certainly, papa has adopted a strange plan to keep us out of the walls of great cities."

"Yes," quietly answered Alexis; "there are not many cities where bears abound."

"Such strange conditions!" added Ivan. "I wonder what father can mean by imposing them upon us."

"Indeed, I hardly know myself. One thing only seems to explain it."

"What is that, brother?"

"You are aware, Ivan, of the interest that papa takes in all matters relating to bears. As people say, it is almost a mania with him."

"O, the great picture in the gallery will account for that," said Ivan, laughing. "But for a *bear*, you know, our papa would never have been a *baron*."

"True: that may have been what first led him to take an interest in these animals."

"And yet to impose upon us these queer conditions!" continued Ivan; "it certainly does seem a little eccentric?"

"No doubt papa has his purpose," said Alexis; "who knows that he may not be intending to write a *monograph* of the bears; and it is for this he wishes to have a full set of their skins, — the complete costume of each individual member of Monsieur Bruin's family? Well, we must do our best, and procure them for him. It is not for us to inquire into the motives of our dear father. It is our duty to obey his orders, even though the task be ever so irksome or difficult."

" O, certainly, brother ! I admit that ; and I am ready to yield obedience and perform any task dear papa may think proper to impose on us."

Certainly there was some reason for the surprise with which the youths had read the letter. Its contents might have appeared still more whimsical to them, had it not been their father that had written it ; and, but for the fact that he had already given them a thorough training in the natural sciences, they would have found it difficult, if not altogether impossible, to carry out his instructions. A bear of every known variety was to be killed and skinned, — killed, too, in its own haunts and by their own hands ; which, of course, meant that they were to visit every country where bears are to be found, and obtain a skin of each kind. Notwithstanding their youth, both boys were skilled hunters, and excellent marksmen. Himself brought up to the calling, their father had early initiated them into the hunter's craft ; and, in addition to the knowledge of natural history which he had imparted, he had taught them habits of self-reliance, — such as are only acquired by ordinary individuals at the full age of manhood. Both were already inured to such perils and hardships as are incidental to a hunter's life ; both could endure to go a day or two without food or drink, — could sleep in the open air, with no other tent than the canopy of heaven, and no other couch than the grassy covering of the earth. All this sort of experience they had already gone through, in the cold climate of their own country ; and it was not likely they would meet with one much more rigorous anywhere on the earth. The young Grodonoffs had been submitted to a

training of almost Spartan severity,—a perfect *Cyropæ-
dia*,—and dreaded neither hardships nor dangers. They
were just the youths to carry out that singular programme
which had been traced for them by the paternal hand.

Was it possible, however, to do so? This was their
first query. There were some very nice points in that
brief chapter of instructions. *Latitudinally* they might
traverse as circumstances required, but not *longitudinally*.
Under these limitations would it be possible to visit the
haunts of all the bears,— to cover, as it were, the whole
area of Bruin's geographical distribution?

That it was possible might be inferred, from the fact
of their father having issued the orders; but it was necessary for the young expeditionists to set out with caution:
else might they take a wrong route, and be altogether
unable to fulfil his injunctions. They must *not twice
cross the same meridian.* It was this quaint condition
that puzzled them, and rendered it necessary to guard
against making a false start.

Lucky it **was** that Alexis was an accomplished zoölo-
gist, and thoroughly understood the geographical distribution of the genus *ursus.* But for this knowledge, they
would certainly have been puzzled as to the route they
were to take.

" Well, brother Ivan !" said Alexis, with a smile, " had
these orders been issued when the great Swede published
his *Systema Naturæ,* our task would have been easily accomplished. How far do you suppose our travels need
to have extended ? "

" I don't quite comprehend you, Alexis. How far ? "

" Why, simply into the court-yard of our palace. It

would have been only necessary to kill and skin one of the great bears chained by the gate, and that would have fulfilled all the conditions papa has imposed upon us."

"And yet, I don't understand you," rejoined Ivan, with a puzzled look.

"How obtuse you are, brother! Read the letter again; note well its terms!"

"Well, I understand them. We are to travel on, and not come home again till we have killed a bear of every variety known."

"There — just so. Of course papa means every variety known to naturalists, — that is, to the 'scientific world,' as it is termed. Now you comprehend my meaning?"

"O, yes!" answered Ivan; "you mean that when Linnæus published his 'System of Nature,' only our own brown bear of Europe was known to naturalists?"

"Precisely so — only the *ursus arctos ;* and consequently we should have had but a very short journey to make, compared with what is before us now. It is true that previous to his death, the Swedish naturalist had made the acquaintance of the 'Polar' bear (*ursus maritimus*) ; but, strange enough, he regarded this as a mere variety of the *ursus arctos* — an error that one may wonder Linnæus could have made."

"O, they are very different. I could tell that myself. To say nothing of the color, they are unlike in shape ; and, as everybody knows, their habits are very dissimilar. Why, one lives in forests, and feeds chiefly upon fruits ; while the other dwells amidst fields of snow and ice, and subsists almost exclusively on flesh, or fish! Variety, indeed! no, they are surely different species."

"Undoubtedly," answered Alexis ; "but we shall have an opportunity of comparing them hereafter. For the present we must drop the subject, and find out the route of travel which papa has traced out for us."

"But he has not indicated any route — has he ? He gives us permission to go where we please, so long as we get the bear-skins, and do not return upon our meridian. We are not to take the *back-track* — ha ! ha ! ha !"

"Of course not ; but you will find, to avoid doing this, we shall have to go by a definite course, and can take no other."

"By my word ! brother, I don't see what you mean. I shall trust all to you : so take me where you please — which way, then ? "

"Ah, that has yet to be determined. I cannot tell myself ; and it will take me some time before I can make quite sure as to what direction we are to take on starting out — whether east, west, north, or south. It will be necessary for me to examine a map of the world, and trace out the boundaries of the different countries in which King Bruin holds sovereignty."

"Ah ! that will be an interesting lesson for me. Here is the map ; let me spread it out, and do what I can to assist you in finding our way."

As Ivan said this, he drew a large travelling map of the world from its case and opening it out, laid it upon the table. Both the youths sat down ; and running their eyes over the chart, proceeded to discuss the direction which, by the conditions imposed upon them, they must necessarily take.

CHAPTER V.

TRACING THE ROUTE.

"In the first place," said Alexis, "there is the brown bear (*ursus arctos*). Him we might find in many parts of our own country, since he is emphatically our 'Russian bear'; but there is also a black bear, which some naturalists say is a variety of the *ursus arctos*, while others believe it to be a separate species, having given to it the specific name of *ursus niger* — *ursus ator* it is sometimes called. Now, whether it be a species or only a variety, we must get a skin of it all the same — since papa has definitely expressed it so."

"This black bear is to be found in our own northern forests, is it not?"

"Yes; it has been observed there; but more frequently in the mountains of Scandanavia; and, as we might wander through all the north of Russia without finding one, our best plan will be to proceed at once to Norway or Lapland. There we shall be certain also of finding the brown bear, and thus kill two birds with one stone."

"Say Lapland: I should like to see the little Laps; but where next? To North America, I suppose?"

"By no means. There is a bear in the Pyrenees, and other mountains of Spain — in the Asturias espe-

cially. It is also deemed by most naturalists to be only
a variety of the *ursus arctos*, but it is certainly a distinct
species ; and papa thinks so. Some naturalists would
have it that there are only three or four distinct species
in the whole world. They might just as well say there
was but one. I think it better to follow papa's views
upon this subject, and regard all those bears which have
permanent marks of distinction — whether it be in size,
color, or otherwise — as being so many separate species,
however much they may approximate in habits or disposition. Why, some naturalists even call the American
black bear merely a variety of our brown ; and, as I said
a moment ago, Linnæus himself believed the Polar to be
the same species. This is now known to be an erroneous
theory. Since papa has given as much time to the study
of the bears as perhaps any one else, I shall follow his
theory, and regard the Spanish bear (*ursus pyrenaicus* it
is called) a distinct species."

"You propose, then, to go next to Spain, and kill the
Spanish bear ? "

"We *must*. Having started in a westerly course by
going to Lapland, we must keep on in that direction."

"But how about the white bear of the Alps ?"

"You mean the *ursus albus* of Lesson ?"

"Yes. To reach the Alps, where it is said to be
found, we should have to recross a meridian of longi-
tude ?"

"We should, if there were such an animal to be found
in the Alps ; but there is not. The white bear of Buffon
and Lesson (*ursus albus*) was only a mere accidental
variety or *albino* of the brown bear ; and papa does not

mean that we should collect the skins of such as these. He has said so. Indeed, Ivan, were that task imposed on us, we should both be old men before we could complete it, and return home again. It is only the skins of the *permanent* varieties we are to procure, and therefore the *ursus albus* is scratched out of our list."

" Out with him then! where go we next? To North America, then, no doubt?"

" No."

" Perhaps to Africa?"

" No."

" Are there no bears in Africa?"

" That is a disputed point among writers, and has been so since the days of Pliny. Bears are mentioned as having been exhibited in the Roman circus, under the name of *Numidian* bears; and Herodotus, Virgil, Juvenal, and Martial, all mention *Lybian* bears in their writings. Pliny, however, stoutly denies that there were any of these animals in Africa; but it must be remembered that he equally denies that stags, goats, and boars existed on the African continent: therefore his statement about the non-existence of the Numidian bears is not worth a straw. Strange enough, the point is as much disputed now as in the days of Pliny. The English traveller, Bruce, states positively that there are no bears in Africa. Another English traveller to Abyssinia, Salt, makes no mention of them; while the German, Ehrenberg, says that he has seen them in the mountains of Abyssinia, and heard of them also in Arabia Felix! Several French and English travellers (Dapper, Shaw, Poncet, and Poiret) bear testimony to the existence of bears in different parts of Africa, — in Nubia, Babur, and Congo. In the Atlas Mountains, between Algiers and Morocco, according to Poiret, bears are common enough; and this writer even gives some details as to their habits. He says that they are exceedingly fierce and carnivorous, and that the Arabs believe they can lift stones in their paws and fling them at those who may be in pursuit of them! He relates that an Arab hunter brought him the skin of one of those bears; and also showed him a wound in his leg, which he had received by the animal having launched a stone at him while he was pursuing it! Monsieur Poriet, however, does not vouch for the truth of the stone-throwing, though he stoutly maintains the existence of African bears."

" What does papa think about it ? " inquired Ivan.

" That there are bears in Africa — perhaps in all the mountainous parts of Africa — but certainly in the Atlas and Tetuan ranges. Indeed, an English traveller of veracity has put the question beyond a doubt, by giving some points in the description of these African bears. Naturalists thought that if such an animal existed in Africa, it would be the same species as the Syrian ; but although the bears reported in the Arabian and Abyssinian mountains are likely enough to be of that species, those of the Atlas are evidently not only distinct from the Syrian bear, but from all other known kinds. One that was killed near Tetuan, about twenty-five miles from the Atlas Mountains, was a female, and less in size than the American black bear. It was black also, or rather brownish black, and without any white marking about the muzzle, but under the belly its fur was of a reddish orange. The hair was shaggy, and four or five inches long, while the snout, toes, and claws were all shorter than in the American black bear, and the body was of thicker and stouter make. The Englishman had learnt something of its habits too. The Arabs said it was rarely met with near Tetuan; that it fed on roots, acorns, and fruits, but was only an indifferent climber. Indeed, it would be very improbable," continued Alexis, " that the great ranges of the Atlas and Abyssinian mountains should be without these mammalia, since they exist in nearly all the other mountains of the globe. Moreover, it should be remembered that it is only a few years since the bears of the Himalayas, of the Great Andes of America, and those of the East Indian Islands — and even the bear of Mount Lebanon — became known to the scientific world. Why, then, should there not be a species in Africa — perhaps more than one — though civilized people are yet unacquainted with it ? "

" But you say we are not going to Africa ? "

" No ; our instructions relate only to every variety of bear known to naturalists ; and the African bear does not come under this category, since it has not yet been described by any naturalist. For that reason we shall have no errand into Africa."

" Then, surely, North America is our next stage ? "

" Certainly not ; you are aware that there is a South American bear."

" Yes, the 'spectacled bear,' as he is called."

" Just so, — the *ursus ornatus*. I think we shall find two species in South America, though that is also a disputed point."

" Well, brother, what if we should ? "

" Why, both will be found in the Andes of Chili and
Peru, and not in the eastern parts of South America."

" And how should that affect our route of travel ? "

" Very essentially, indeed. Were we to go first to
North America, we should find no less than five species,
or four species and one well-marked variety. To reach
the native haunt of one of these — I mean the grisly
(*ursus ferox*) — we should have to go farther west than
any part of the South American Andes : how, then,
could we afterwards reach the spectacled bear without
doubling back on our meridian ? "

" True, brother ; I see that, by looking on the map.
You propose, then, steering first to South America, and
afterwards to the northern division of the American
continent ? "

" We are compelled to do so, by the very nature of
our contract. Having procured the skins of *ursus ornatus* and another variety we shall find in the Andes, we
can then travel almost due north. On the Mississippi we
shall be able to pick up a skin of the American black
bear (*ursus americanus*), and by the help of the Hudson's
Bay *voyageurs* we shall reach the shores of the great gulf
from which that territory takes its name. There the
' polar bear ' (*ursus maritimus*) can be found. Farther
westward and northward we may hope to capture the
' barren-ground bear,' which the English traveller, Sir
John Richardson, thinks is only a variety of our European brown bear, but which papa — and good reasons he
has — believes to be nothing of the kind. Crossing the
Rocky Mountains, we shall be able, I hope, to knock over

the famed and formidable grisly (*ursus ferox*), and in Oregon, or British Columbia, we shall strip his hide from the 'cinnamon bear' (*ursus cinnamonus*), believed to be a variety of the American black. That will finish with the bears of America."

"Asia next, I suppose?"

"Yes, straight across to Kamschatka. There we shall meet with the 'Siberian,' or 'collared bear' (*ursus collaris*). Of these, two varieties are said to exist, one of which, specified by the name **ursus sibiricus**, is also found in Lapland and Siberia."

"Go on, brother! Where next?"

"From Kamschatka we shall make a long traverse to the southwest. Our next hunting-ground will be Borneo."

"Ah! the beautiful little bear with the orange-colored breast!"

"Yes; that is the 'Bornean bear' (*ursus euryspilus*) or 'Bruang,' as he is called by the Malays."

"But there is another Bruang?"

"Yes, — the 'Malayan sun bear' (*ursus malaganus*). This we shall encounter in Sumatra or Java, whichever we choose to visit."

"Well, the list is much larger than I expected: certainly it has been wonderfully lengthened since the days of the good old Linnæus."

"We have not reached the end yet."

"Where next, brother?"

"Up the Bay of Bengal, and on to the Himalayas. First in the foot hills of these mountains we shall have to search for the curious 'sloth bear,' or 'juggler's bear' (*ours de jongleurs*) as the French writers term him.

He is the *ursus labiatus* of naturalists; and we may find him in the plains of India, before reaching the Himalayas. Having skinned him, we shall proceed to climb the great mountains, and higher up we are certain to come across the '**Thibet** bear' (*ursus thibetanus*) — by some very erroneously described as being one of the numerous varieties of the European brown bear! Still higher up we shall, I hope, have the good luck to encounter and kill a specimen of the 'Isabella bear' (*ursus isabelinus*), so called from his color, but termed by Anglo-Indian **sportsmen the** 'snow bear,' because he **frequents the** declivities **near the** snow-line of these stupendous mountains."

"That is all, is it not?"

"No, Ivan, — one more, and that will be the **last**."

"What is he?"

"The 'Syrian' (*ursus syriacus*); and though the last in our catalogue, this is the very first on record: for they were bears of this species that came out of the wood and 'tare forty and two' of the mockers of the prophet Elisha. We shall have to visit Syria to procure a skin of the *ursus syriacus*."

"Well, I hope their ferociousness has been tamed down since Elisha's time, else we may stand a fair chance of being served in a similar fashion."

"No doubt **we** shall have many a scratch before **we** encounter the bears of Mount Lebanon. When we have obtained a robe from one of them, there will be nothing more for us to do but take the most direct route home. We shall then have gone *once round the world*."

"Ah, that we shall!" said Ivan, laughing; "and all

over it too. Great Czar! I think by the time we have captured one of Elisha's bears we shall have had a surfeit of travel."

"No doubt of it; but now, brother, that we know where we are going, let us waste no more time, but signify our acceptance of the conditions, and be off at once."

"Agreed," said Ivan; and both returning into the presence of the baron, announced their readiness to take the road.

"Are we to travel alone, papa?" inquired Ivan; "I think you spoke of an attendant?"

"Yes, one attendant. You must not be encumbered with too many servants to wait upon you." One will be quite sufficient."

"Who is it to be?" asked Ivan.

The baron rang the bell, and a servant entered.

"Send Corporal Pouchskin to me!"

Shortly after, the door reopened, and a man of about fifty appeared. The tall, well-balanced form and erect attitude — the close-cropped hair and enormous grizzled moustache, combined with great gravity of features, denoted a veteran of the Imperial Guard, — one of those grand and redoubtable soldiers who have seen service in the presence of an emperor. Though no longer wearing the military uniform, but dressed somewhat as a park or game-keeper, the silent salute and attitude of "attention" were sufficiently indicative of the profession which Pouchskin had followed: for it was the veritable Pouchskin who had entered the apartment. He said not a word, nor did he look either to the right or left, — only directly forward, and at the baron.

" Corporal Pouchskin ! "

" General ! "

" I wish you to make a **journey.**"

" I am ready."

" Not quite, corporal. I will give **you an hour to** prepare."

" Where does the general wish me to go ? "

" Round the world."

" Half an hour will suffice."

" So much the better, then. Prepare to start in half an hour."

Pouchskin bowed and retired.

CHAPTER VI.

TO THE TORNEA.

WE shall not detail the parting interview between the Baron Grodonoff and his sons; there was the usual interchange of affectionate expressions, with as much feeling as is common on such occasions. Neither need we relate the ordinary incidents of travel which befell our expeditionists on their way to the mountains of Lapland. Suffice it to say that they journeyed by post from St. Petersburg direct to Tornea, at the head of the Great Bothnian Gulf. Thence they proceeded northward up the river Tornea — till they had reached the mountainous region in which this stream takes its rise. They were amply furnished with the means of travelling in the most expeditious manner, and were not encumbered with any great amount of luggage. A bag of roubles, which Pouchskin carried in a safe pocket, proved the most convenient article they could have taken along with them; since it enabled them to supply their wants from day to day, without troubling themselves with any cumbersome baggage. There are few parts of the world in which ready money will not command the necessaries of life; and as this was all our hunters cared for, they had no difficulty in obtaining supplies — even in the remote regions of uncivilized Lapland. The wild, half-savage Lap perfectly comprehends the value of a coin; and will exchange for it his reindeer flesh and milk, or anything else that may be asked from him. Our young hunters, therefore, travelled lightly — with little else in the shape of baggage than a pair of knapsacks which they carried on their backs, and which contained only a change or two of linen, and such toilet articles as were absolutely necessary to their comfort. A knapsack of much larger dimensions formed the chief care of Pouchskin; and although this, with its contents, would have been a heavy load for an ordinary man, the veteran of the Imperial Guard thought no more of it than if it had been a bag of feathers. Each in addition carried an ample fur cloak; which, on the march, was folded up and strapped to their backs on top of the knapsack, but at night was wrapped around their bodies, and served both as bed and bedclothes. All three were armed and equipped, in the most substantial manner. They carried guns, though differing in kind. The piece of Alexis was a handsome Jäger rifle; Ivan's was a double-barrelled shot-gun or fowling-piece; while Pouchskin balanced over his shoulder an immense fusil, the bullet of which weighed a good ounce avoirdupois. All were provided with a knife of one fashion or another.

In such guise did our young hunters enter the mountains of Lapland; and commence their search after the "old man in the fur coat," as the Laplanders term the bear.

They had taken proper measures to secure success. They had secured the services of a guide, who engaged to conduct them to a district where bears existed in great plenty, and where he himself lived in a state almost as savage as the bears — for he was a true Laplander, and lived in a tent in the very heart of the mountains. He was one of those who had no reindeer; and was therefore forced to depend on the chase for his subsistence. He trapped the ermine and beaver — killed the wild reindeer when he could — spent his whole life in battling with wolves and bears; and with the skins of these animals — which he sold to the fur-traders — he was able to supply himself with the few necessaries which such a state of existence called for.

Under his tent of coarse *wadmal* cloth the travellers found shelter, and such rude hospitality as the poor Lap could afford them — in return for which they had to live in the midst of a smoke that nearly put out their eyes. But they knew they had entered upon an expedition, in which many hardships were to be expected; and they bore the inconvenience with becoming fortitude.

It is not my intention to give the details of the everyday life of the young hunters, nor yet an account of the very many curious incidents, which occurred to them during their sojourn in Lapland. Much was noted down in their journal — from which this narrative has been drawn — interesting only to themselves, or perhaps still more to their father the baron. For him they wrote an account of everything peculiar that they observed — such as the odd customs of the Laplanders — their mode of travelling in sledges with reindeer — their snow-skating on the *skidors* and *skabargers* — and, in short, a full account of the habits and manners of these singular people. Especially, however, did Alexis describe the objects of

natural history which came under his notice — giving
such details as he drew from personal observation, or
derived from the native hunters, many of whom they
encountered while engaged in the chase of the bear.

These details, were they given in full, would fill a book
of themselves. We must content ourselves, therefore,
with relating only the more interesting incidents, and
striking adventures which happened to our heroes.

We may here state that it was in the early part of
spring that they arrived in Lapland, or rather in the
latter part of winter, when the ground is still covered
with deep snow. At this season the bears are hidden
away in their caves — in crevices of the rocks or hollow
trees — from which they only issue forth when the spring
sun makes itself felt, and the snow begins to disappear
from the sides of the hills.

Every one has heard of this *winter sleep* of the bears ;
and it has been attributed to bears of all species. This,
however, is a mistake, as it is only indulged in by a few
kinds ; and the climate and nature of the country which
the bear inhabits has more to do with his *hybernation*
than any natural instinct of the animal ; since it has been
observed that bears will go to sleep, or *hybernate*, as it is
termed, in one part of a country, while individuals of the
same species, in another region, will be found roaming
about all the winter through. The state of torpor seems
to be voluntary with these animals : since it is generally
in districts where food could not be procured that they
submit themselves to this prolonged *siesta*.

However this may be, the brown bears of Lapland
certainly indulge in a period of slumber — during which

they are difficult to find. Never issuing from their places of concealment, they make no track in the snow by which they might be followed. At such seasons it is only by accident, or by the aid of his dog, that the Lap hunter chances to discover the retreat of a bear; and, when one is thus discovered, various methods are adopted for securing the valuable skin and carcass of the animal.

It so chanced that, previous to the arrival of the young Russians upon their hunting-ground, there had been a show of spring — that is, a few days of warm sun — but this had been succeeded by a return of the cold weather, with a fresh fall of snow. The spell of warmth, however, had aroused many bears from their lethargy — some of which had ventured out of their caves, and made short excursions among the hills — in search, no doubt, of the *berries*, that, preserved all winter by the snow, are sweet and mellow at this season, and a favorite food of the bears.

This casual occurrence of the spring having made a promise and not kept it, was just the chance for our hunters; since it enabled them in a very short time to track a bear to his den.

A few days after their arrival upon the hunting-ground, they were able to do this — having come upon the footmarks of a bear, that, followed for a mile or so through the snow, led them to the animal's lair. It led them also to an adventure, which was the first they had yet encountered; and which came very near being the last that Pouchskin was ever to have in the world Pouchskin was certainly in great peril; and how he escaped from it will be learnt, by reading an account of the adventure.

CHAPTER VII.

JACK-IN-THE-BOX.

IT was early in the morning, shortly after leaving the
the tent of the Laplander, they had chanced upon the
track of the bear.

After following it for nearly a mile, it conducted them
to a narrow gorge or ravine, lying between two rocky
ridges. The ravine itself was not more than ten or a
dozen yards in width, and its bottom was filled with
snow to the depth of several feet. Along the sides the
snow lay sparsely; and in fact there had been scarce
any in that place before the fall the preceding night.
This had only covered the ground to the depth of a few
inches: but it was sufficient to show the footmarks of
the bear; and they were able to follow the *spür* — so
the Scandinavian hunters call the tracks of an animal —
as fast as they chose to go.

Following it up, then, our hunters entered the ravine.
They kept for some distance along one side — just by
the edge of the deep snow; but at length the track indicated where the bear had crossed to the other side;
and of course they were compelled to cross likewise.

This deep snow was the accummulated deposits of
different storms that had occurred during the winter;

and, shadowed from the sun by the long branches of evergreen pines from both sides stretching outward over the ravine, it had remained without melting. There was a crust over it — strong enough to carry a man on *skidors*, but not without them, unless he proceeded with care and caution. The bear had gone over it; but these animals, notwithstanding their enormous weight and bulk, can pass over ice or crusted snow that will not carry a man. Their weight rests upon four points instead of two; and as they need only lift one foot at a time, they still have three points of support. A man must also lift one foot, which leaves him only one to stand upon; and therefore his whole weight presses upon a single point, and so endangers his breaking through. The great length of a bear's body, moreover, and the vast stretch between his fore and hind legs give him an additional advantage — enabling him to distribute his weight over a large surface — and this is why he can shuffle over ice or snow-crust that may be too weak to carry a human being. Every boy knows — at least every boy who has skated or ventured upon a frozen pond — that by creeping on hands and knees, or, more certain still, by sprawling along on the breast, ice may be passed over, that would not bear the same boy in an erect attitude.

Such advantage, then, had the bear which our young hunters were tracking up; and it would have been well for them — at least for Pouchskin — had they thought of it. They did not. They supposed that where a great heavy animal like a bear had gone they might go too; and, without further reflection, they stepped out upon the deep bed of snow.

Alexis and Ivan being light weights passed over the
snow safely enough ; but Pouchskin, weighing nearly as
much as both of them — and further loaded with a ponderous wood-axe and his huge gun, to say nothing of sun-
dry well-filled pockets and pouches — was more than the
crust would carry. Just when he had got about half way
across, there was heard a tearing crash ; and before the
boys could turn round to inquire the cause, Pouchskin
had disappeared, and all his *paraphernalia* along with
him !

No, not quite all. There was seen about two feet of
the barrel of his gun above the surface ; and as that still
pointed upward — while it moved around the circular
hole through which the old guardsman had fallen — the
boys concluded that the piece was in his hands, and that
Pouchskin was still upon his feet.

At the same instant a voice reached their ears — in a
hollow sepulchral tone, like that of a man speaking from
the bottom of a well, or through the bunghole of an
empty cask !

Notwithstanding its *baritone* notes, the boys perceived
that the exclamations made by the voice were not those
of terror, but rather of surprise, followed by a slight
laugh. Of course, therefore, their attendant had received
no injury, nor **was** he in any danger ; and, assured of
this, Ivan, first, and then Alexis, broke out into yells of
laughter.

On cautiously approaching the trap-like hole, through
which Pouchskin had disappeared, their merriment burst
forth afresh, at the ludicrous spectacle. There stood the
old guardsman, like a jack-in-the-box in the centre of a

hollow funnel-shaped cylinder which he had made in the snow. But what was strangest of all, there was no snow among his feet: on the contrary, he was up to his knees in water, and not stagnant water' either, but a current, that ran rapidly underneath the snow, and had swished the crusted fragments from the spot where he was standing!

A stream, in fact, ran down the ravine; and, although the snow completely hid it from view, there it was, rushing along underneath through a tunnel which it had melted out for itself — the snow forming a continuous bridge above it.

The boys did not know all this — for they could only just see the top of Pouchskin's head, with his long arms holding the gun — but they could hear the rushing noise of the water, and Pouchskin reported the rest.

It did not appear so easy to extricate him from his unpleasant predicament; for the resemblance between his situation, and that of jack-in-the-box, went no further. There was no jerking machinery by which the ex-guardsman could be jumped out of his box; and, since his head was full three feet below the crust of the snow, how he was to be raised to the surface required some consideration.

Neither of the young hunters dared to approach the circumference of the circular hole through which Pouchskin had sunk. They might have broken through themselves, and then all three would have been in the same fix. Of course, under this apprehension, they dared not go near enough to pull him out with their hands — even had they been able to reach down to him.

It is true he might have got out, after some time, by

breaking the snow before him, and working his way at right angles to the course of the stream : for it was evident that the ground sloped sharply up in that direction, and the snow became shallower. Except above the water, it was firm enough to have borne his weight, and after a time he might have scrambled out; but a more expeditious plan of relieving him, and one far less troublesome to Pouchskin, suggested itself to Alexis.

One of the *impedimenta*, which the old guardsman carried on his shoulders, was a coil of stout cord — almost a rope. This they had brought with them, in the anticipation of being successful in their hunt; and, with the idea of its being required at the skinning of the bear — as also for packing the hide, or any similar purpose.

It was the presence of this cord that suggested to Alexis the scheme he had conceived for relieving his faithful follower from his unhappy position ; and the plan itself will be understood by our describing its execution, which took place on the instant.

Alexis called to Pouchskin to tie one end of the rope round his body, and then fling the other out upon the snow — as far as he could cast it. This request was instantly complied with ; and the end of the rope made its appearance at the feet of Alexis.

The latter taking it in his hand, ran up the bank to the nearest tree ; and giving it a turn or two round the trunk, he handed it to Ivan, with the direction to hold it fast and keep it from slipping. A knot would have served the same purpose ; but the whole thing was the work of only a few moments ; and as Ivan was standing by doing nothing, his brother thought he might just as well take hold of the rope and save time.

Alexis now crept back, as near to the edge of the trap as it was safe to go. He took with him a long pole, which by a lucky chance he had found lying under the trees. Slipping this under the rope, and placing it cross-wise, he shoved it still nearer to the circumference of the broken circle — his object being to give support to the cord, and keep it from cutting into the snow.

The contrivance was perfectly correct; and as soon as Alexis had got all ready, he shouted to Pouchskin to haul upon the rope, and help himself.

Meanwhile, the old guardsman had slung his fusil upon his back; and, immediately on receiving the signal, com-menced his ascent — pulling hand over hand upon the rope, and assisting his arms by working his feet against the wall of snow.

The moment his head appeared above the surface, the laughter of his young masters, that had been for a while suspended, burst forth afresh. And it was no wonder: for the expression upon the old soldier's visage, as it rose above the white crust, his bent attitude, and the desperate exertions he was making to clamber upward, all combined to form a most ludicrous picture.

Ivan screamed till the tears ran down his cheeks. So overcome was he with mirth, that it is possible he would have let go, and permitted Pouchskin to tumble back in-to his trap; but the more sober Alexis, foreseeing such a contingency, ran up and took hold of the rope.

By this means, Pouchskin was at length landed safely on the surface of the snow; but even his tall boots of

Russia leather had not saved his legs and feet from
getting well soaked ; and he was now dripping with
muddy water from the thighs downwards.

There was no time, however to kindle a fire **and dry**
him. They did not think of such **a** thing. **So** eager
were all three in the chase of the bear, that they only
waited to coil up the cord, and then continued onward.

CHAPTER VIII.

THE SCANDINAVIAN BEARS.

"REALLY, now," said Ivan, pointing to one of the tracks, "if it was n't that I see the marks of claws instead of toes, I should fancy we were tracking a man instead of a bear — some barefooted Laplander, for instance. How very like these tracks are to those of a human foot!"

"That is quite true," rejoined Alexis; "there is a very remarkable resemblance between the footprints of the bear and those of a human being — especially when the tracks have stood a while. As it is, now, you can see clearly the marks of the claws; but in a day or two, when the sun or the rain has fallen upon the snow, and melted it a little, the claw-marks will then be filled up with the thaw, and, losing their sharp outlines, will look much more like the tracks of toes. For that reason, an old bear-track is, indeed, as you say, very like that of a human foot."

"And quite as large, too?"

"Quite as large: the tracks of some kinds even larger than those of most men. As, for instance, the white and grisly species — many individuals of both having paws over twelve inches in length!"

"The bear does not tread upon his toes in walking, but

lays the whole sole of his foot along the ground — does he not?" asked Ivan.

"Precisely so," replied Alexis; "and hence he is termed a *plantigrade* animal, to distinguish him from those other kinds, as horses, oxen, swine, dogs, cats, and so forth, that all, in reality, step upon their toes."

"There are some other plantigrade animals besides bears?" said Ivan, interrogatively; "our badger and glutton, for instance?"

"Yes," answered the naturalist. "These are plantigrade; and for this reason they have been classed along with the bears under the general name *ursidæ;* but in father's opinion, and mine too," added Alexis, with a slight sparkle of scientific conceit, "this classification is altogether an erroneous one, and rests upon the very insignificant support of the plantigrade feet. In all other respects the different genera of small animals, that have thus been introduced into the family of the bears are as unlike the latter almost as bears are to bluebottles."

"What animals have been included in this family *ursidæ?*"

"The European glutton and American wolverene (*gulo*), the badgers of both continents, and of Asia (*meles*), the raccoon (*procyon*), the Cape ratel (*mellivora*), the panda (*ailurus*), the benturong (*ictides*), the coati (*nasua*), the paradoxure (*paradoxurus*), and even the curious little teledu of Java (*mydaus*). It was Linnæus himself who first entered these animals under the heading of *bears* — at least, such of them as were known in his day; and the French anatomist, Cuvier, extended this incongruous list to the others. To distinguish them from the true bears, they divided the family into two branches — the *ursinæ*, or bears properly so called, and the *subursinæ*, or little bears. Now, in my opinion," continued Alexis, " there is not the slightest necessity for calling these numerous species of animals even '*little bears.*' They are not bears in any sense of the word : having scarce any other resemblance to the noble Bruin than their plantigrade feet. All these animals — the Javanese teledu excepted — have long tails; some of them, in fact, being very long and very bushy — a characteristic altogether wanting to the bears, that can hardly be said to have tails at all. But there are other peculiarities that still more widely separate the bears from the so-called 'little bears;' and, indeed, so many essential points of difference, that the fact of their being classed together might easily be shown to be little better than mere anatomical nonsense. It is an outrage upon common sense," continued Alexis, warming with his subject, " to regard a raccoon as a bear, — an animal that is ten times more like a fox, and certainly far nearer to the genus *canis* than that of *ursus.* On the other hand, it is equally absurd to break up the true bears into different *genera* — as these same anatomists have done ; for if there be a family in the world the individual members of which bear a close family likeness to one another, that is the family of Master Bruin. Indeed, so like are the different species, that other learned anatomists have gone to the opposite extreme of absurdity, and asserted that they are all one and the same ! However, we shall see as we become acquainted with the different members of this distinguished family, in what respects they differ from each other, and in what they are alike."

" I have heard," said Ivan, " that here, in Norway and Lapland, there are two distinct species of the brown bear, besides the black variety, which is so rare ; and I have also heard say that the hunters sometimes capture a variety of a grayish color, which they call the ' silver bear.' I think papa mentioned these facts."

" Just so," replied Alexis ; " it has been the belief among Swedish naturalists that there are two species, or at least permanent varieties, of the brown bear in Northern Europe. They have even gone so far as to **give** them separate specific names. **One** is the *ursus arctos major*, while the other is *ursus arctos minor*. The former is the larger animal — more fierce in its nature, and more carnivorous in its food. The other, or smaller kind, is of a gentler disposition — or at all events more timid — and instead of preying upon oxen and other domestic animals, confines itself to eating grubs, ants, roots, berries, and vegetable substances. In their color there is no perceptible difference between the two supposed varieties, more than may be often found between two individuals notedly of **the** same **kind ; and** it is only in size and habits that a distinction has been observed. The latest and most accurate writers upon this subject believe that the great and little brown bears are not even varieties ; and that the distinctive characteristics **are** merely the effects of age, sex, or other accidental circumstances. It is but natural to suppose that the younger bears would not be so carnivorous as those of greater age. **It is** well known that preying upon other animals and feeding upon their flesh, is not a natural instinct of the brown bear ; it **is** a habit that has its origin, first, in the scarcity of other food, but which, once entered upon, soon develops itself into a strong propensity, — almost equalling that of the *felidæ*.

"As to the black bear being a distinct species, that is a question also much debated among both hunters and naturalists. The hunters say that the fur of the black European bear is never of that jetty blackness which characterizes the real black bears of American and Asiatic countries, but only a very dark shade of brown; and they believe that it is nothing more than the brown fur itself, grown darker in old age. Certainly they have reason for this belief: since it is a well-known fact that the brown bears do become darker as they grow older.

"Ha!" said Ivan, with a laugh, "that is just the reverse with us. Look at Pouchskin there! Your hair was once black, was n't it, old Pouchy?"

"Yes, Master Ivan, black as a crow's feathers."

"And now you 're as gray as a badger. Some day, before long — before we get home again, maybe — your moustache, old fellow, will be as white as an ermine."

"Very like, master, very like, — we 'll all be a bit older by that time."

"Ha! ha! ha!" laughed Ivan; "you 're right there, Pouchy; but go on brother!" he added, turning to Alexis; "let us hear all about these Scandinavian bears. You have not spoken of the 'silver' ones."

"No," said Alexis; "nor of another kind that is found in these countries, and that some naturalists have elevated into a different species — the 'ringed bear.'"

"You mean the bears with a white ring round their necks? Yes, I have heard of them too."

3   D

"Just so," rejoined Alexis.

"Well, brother, what do *you* think? Is it a distinct species, or a permanent variety?"

"Neither one nor the other. It is merely an accidental marking which some young individuals of the brown bear chance to have, and it scarcely ever remains beyond the age of *cubhood*. It is only very young bears that are met with of this color; and the white ring disappears as they get older. It is true that hunters now and then meet with an odd ringed bear of tolerable size and age; but all agree that he is the brown bear, and not a distinct kind. The same remarks apply to the 'silver' bear; and hunters say that in a litter of three cubs they have found all three colors — the common brown, the 'ringed,' and the 'silver,' — while the old mother herself was a true *ursus arctos*."

"Well, since papa only binds us to the brown and black, it will be a nice thing if we could fall in with a skin of the ringed and silver varieties. It would please him all the better. I wonder now what sort is this fellow we are following? By the size of his tracks he must be a whopper!"

"No doubt an old male," rejoined Alexis; "but if I am not mistaken, we shall soon be able to determine that point. The *spär* gets fresher and fresher. He must have passed here but a very short while ago; and I should not wonder if we were to find him in this very ravine."

"See!" exclaimed Ivan, whose eyes had been lifted from the trail, and bent impatiently forward; — "see! by the great Peter! yonder's a hole, under the root of that tree. Why might it not be his cave?"

"It looks like enough. Hush! let us keep to the trail, and go up to it with caution — not a word!"

All three, now scarce breathing — lest the sound should be heard — stole silently along the trail. The fresh-fallen snow, still soft as eider-down, enabled them to proceed without making the slightest noise; and without making any, they crept up, till within half a dozen paces of the tree.

Ivan's conjecture was likely to prove correct. There was a line of tracks leading up the bank; and around the orifice of the cavity the snow was considerably trampled down — as if the bear had turned himself two or three times before entering. That he had entered, the hunters did not entertain a doubt: there were no return tracks visible in the snow — only the single line that led up to the mouth of the cave, and this seemed to prove conclusively that Bruin was "at home."

CHAPTER IX.

HYBERNATION OF BEARS.

As already stated, it is the custom of the brown bear, as well as of several other species, to go to sleep for a period of several months every winter, — in other words, to *hybernate*. When about to take this long nap, the bear seeks for himself a cave or den, in which he makes his bed with such soft substances as may be most convenient — dry leaves, grass, moss, or rushes. He collects no great store of these, however — his thick matted fur serving him alike **for bed** and coverlet; and very often he makes no further ado about the matter than to creep into the hole he has chosen, **lie** down, snugly couch his head among the thickets **of long hair that cover his** hams, and thus go to sleep.

Some naturalists have asserted that this sleep is a state **of** torpidity — from which the animal **is** incapable of awaking himself or of being awakened, until the regular period of indulgence in it may have passed. This, however, is not the **case**; for bears are often surprised in their sleep, and when aroused by the hunters act just as is usual with them at other times.

It must be observed, however, that the retirement of **the** bear into winter quarters is not to be regarded as of

the same nature as the hybernation of marmots, squirrels, and other species of rodent animals. These creatures merely shut themselves up from the cold; and to
meet the exigencies of their voluntary imprisonment,
they have already collected in their cells a large store of
their usual food. Bees and many other insects do precisely the same thing. Not so with the bear. Whether
it be that he is not gifted with an instinct of providence it
is difficult to say; but certain it is, that he lays up no
store for these long dark days, but goes to sleep without
thought of the morrow.

How he is maintained for several months without eating is one of nature's mysteries. Every one has heard
the absurd theory, that he does so by "sucking his
paws," and the ingenious Buffon has not only given cre-
dence to this story, but endeavors to support it, by stating
that the paws when cut open yield a substance of a
milky nature!

It is a curious fact that this story is to be found scattered all over the world — wherever bears hybernate.
The people of Kamschatka have it; so also the Indians,
and Esquimaux of the Hudson's Bay territory, and the
Norwegian and Lap hunters of Europe. Whence did
these widely-distributed races of men derive this common idea of a habit which, if the story be a true one,
must be common to bears of very different species?

This question can be answered. In northern Europe
the idea first originated — among the hunters of Scandinavia. But the odd story once told was too good to be
lost; and every traveller, since the first teller of it, has
taken care to embellish his narrative about bears with

this self-same conceit; so that, like the tale of the Amazon women in South America, the natives have learnt it from the travellers, and not the travellers from the natives!

How absurd to suppose that a huge quadruped, whose daily food would be several pounds weight of animal or vegetable matter — a bear who can devour the carcass of a calf at a single meal — could possibly subsist for two months on the *paw-milk* which M. Buffon has described!

How then can we account for his keeping alive? There need be no difficulty in doing so. It is quite possible that during this long sleep the digestive power or process is suspended, or only carried on at a rate infinitesimally small; that, moreover, life is sustained and the blood kept in action by means of the large amount of fat which the bear has collected previous to his *going to bed*. It is certain that, just at their annual *bedtime*, bears are fatter than at any other season of the year. The ripening of the forest fruits, and the falling of various seeds of mastworts, upon which, during the autumn, bears principally subsist, then supply them with abundance, and nothing hinders them to get fat and go to sleep upon it. They would have no object in keeping awake: were they to do so, in those countries where they practise hybernation, they would certainly starve, for, the ground being then frozen hard, they could not dig for roots, and under the deep covering of snow they might search in vain for their masts and berries. As to foraging on birds or other quadrupeds, bears are not fitted for that. They are not agile enough for such a purpose.

They will eat both when they can catch them; but

they cannot always catch them; and if they had no other resource in the snowy season the bears would certainly starve. To provide them against this time of scarcity, nature has furnished them with the singular power of somnolence. Indeed, that this is the purpose is easily proved. It is proved by the simple fact that those bears belonging to warm latitudes, as the Bornean, Malayan, and even the black American of the Southern States do not hybernate at all. There is no need for them to do so. Their unfrozen forests furnish them with food all the year round; and all the year round are they seen roaming about in search of it. Even in the Arctic lands the polar bear keeps afoot all the year; his diet not being vegetable, and therefore not snowed up in winter. The female of this species hides herself away; but that is done for another purpose, and not merely to save herself from starvation.

That the stock of fat, which the bear lays in before going to sleep, has something to do with subsisting him, is very evident from the fact that it is all gone by the time he awakes. Then, or shortly afterwards, Master Bruin finds himself as thin as a rail; and were he to look in a glass just then, he would scarce recognize himself, so very different is his long emaciated carcass from that huge, plump, round body, that two months before he could scarce squeeze through the entrance to his cave!

Another great change comes over him during his prolonged sleep. On going to bed, he is not only very fat, but also very lazy; so much so that the merest tyro of a hunter can then circumvent and slay him. Naturally a well-disposed animal — we are speaking only of the

brown bear (*ursus arctos*) though the remark will hold
good of several other species — he is at this period more
than usually civil and soft-tempered.   He has found a
sufficiency of vegetable food, which is more congenial to
his taste than animal substances; and he will not molest
living creature just then, if living creature will only let him
alone.   Aroused from his sleep, however, he shows a different disposition. He appears as if he had got up "wrong
side foremost." His head aches, his belly hungers, and
he is disposed to believe that some one has stolen upon
him while asleep, and robbed him of his suet.   Under
this impression he issues from his dark chamber in very
ill-humor, indeed. This disposition clings to him for a
length of time; and if at this period, during his morning
rambles, he should encounter any one who does not get
speedily out of his way, the party thus meeting him will
find him a very awkward customer. It is then that he
makes havoc among the flocks and herds of the Scandinavian shepherd — for he actually does commit such ravages — and even the hunter who meets him at this
season will do well to "ware bear."

And so does the hunter; and so did Alexis, and Ivan,
and Pouchskin. All three of them were well enough
acquainted with the habits of the bear — their own Russian bear — to know that they should act with caution in
approaching him.

And in this wise they acted; for, instead of rushing up
to the mouth of the hole, and making a great riot, they
stole forward in perfect silence, each holding his gun
cocked, and ready to give Bruin a salute the moment
he should show his snout beyond the portals of his den.

Had they not tracked him to his cave, they would have acted quite differently. Had they found a bear's den — within which they knew that the animal was indulging in his winter sleep — they would not have cared so much how they approached it. Then he would have required a good deal of stirring up to induce him to show himself, so that they could get a shot at him ; but the track told them that this one had been up and abroad — perhaps for several days — and as the new snow, in all likelihood, had hindered him from picking up much to eat, he would be as " savage as a meat-axe."

Expecting him to spring out almost on the instant, the three took stand at some distance from the mouth of the cave ; and, with arms in readiness, awaited his coming forth.

3 *

CHAPTER X.

BRUIN AT HOME?

THE entrance to the cave, if cave it was, was an aperture of no great dimensions — about large enough to admit the body of a full-grown bear, and no bigger. It appeared to be a hole or burrow, rather than a cave, and ran under a great pine-tree, among whose roots, no doubt, was the den of the bear. The tree itself grew up out of the sloping bank; and its great rhizomes stretched **over** a large space, many of them appearing above the surface soil. In front of the aperture was a little ledge, where the snow was hacked by the bear's paws, but below this ledge the bank trended steeply down — its slope terminating in the bed of deeper snow already described.

As stated, the three hunters had taken their stand, but not all together. Directly in front of the cave was Pouchskin, and below it, of course, on account of the sloping bank. He was some six paces from the aperture. On the right side Ivan had been placed, while Alexis had passed on, and now stood upon the left. The three formed a sort of isosceles triangle, of which Pouchskin was the apex, and the line of the bank the base. A perpendicular dropped from the muzzle of Pouchskin's gun would have entered the aperture of the cave. Of course

Pouchskin's was the post of danger; but that was to be expected.

They stood a good while in silence. No signs of Bruin — neither by sight nor hearing.

It was then resolved that some stir should be made — a noise of any kind, that might bring the beast forth. They coughed and talked loudly, but all to no purpose. They shouted at length with like fruitless result — Bruin would not stir!

That he was inside none of them doubted. How could they? The tracks going to the cave, and none coming from it, set that question at rest. Certainly he was in his den? but whether asleep or not, it was evident he took no heed of their shouting.

Some other means must be adopted to get him out. He must be stirred up with a pole! This was the plan that suggested itself, and the one determined upon.

Pouchskin started off to procure a pole. The others kept guard — still holding their guns in readiness, lest the bear might make a rush in Pouchskin's absence. But Bruin had no such intention; nor was his presence betrayed by sight or sound, until Pouchskin came back. He had cut a pole with his axe, and had taken the precaution to select a long one. A young sapling it was, that when cleared of its branches appeared as long as a hop-pole. Pouchskin knew the advantage of its length. He had no particular wish to come to close quarters with the bear.

Creeping back pretty nearly into his old place, he inserted the end of the sapling into the aperture — then rattled it against the sides, and waited a bit.

No response from Bruin !

Once more the pole was pushed in, this time a little further, and again accompanied with similar noisy demonstrations.   Bruin neither moves nor makes sound !

"He must be asleep! Try a little further, Pouchskin!"

This suggestion came from the impatient Ivan.

Encouraged by the words of his young master, Pouchskin approached nearer to the aperture, and buried half of the pole inside.   He then turned the stick and poked it all about, but could touch nothing that felt like a bear. Growing more confident, he crept yet nearer, and pushed the pole up till he could touch the bottom of the cave — once more feeling with its point in all directions, against the further end, along the sides, upwards and downwards, and everywhere.   Still he touched nothing soft — nothing that felt as the shaggy hide of a bear should do — nothing, in fact, but hard rocks, against which the stick could be heard rattling wherever he pushed it !

This was very mysterious.   Pouchskin was an old bear-hunter.   He had poked his pole into many a burrow of Bruin, and he knew well enough when he had touched bottom.   He could tell moreover that the cave he was now exploring was all in one piece — a single-roomed house.   Had there been any second or inner chamber he would have found the aperture that led to it ; but there appeared to be none.

To make sure of this, he now approached quite near to the entrance, and continued to gauge the cavity with his stick.   Alexis and Ivan also drew near — one on each side of him — and the exploration continued.

In a short while, however, Pouchskin became nearly satisfied that *there was no bear in the den!* He had groped with his stick all round and round it, and had come in contact with nothing softer than a rock or a root of the tree. As a last *resource* he lay down on the ground to listen, placing his ear close to the mouth of the cave; and, cautioning his young masters to keep silent, in this position he remained for some seconds of time.

Perhaps it was fortunate for them, if not for him, that they attended to his caution. Their silence enabled them to hear what Pouchskin could not — placed as he now was — and that was a sound that caused the young bear-hunters to start back and look upwards, instead of into the cave.

As they did so, a sight met their eyes that drew from both a simultaneous cry, while both at the same instant retreated several paces from the spot, elevating their guns as they went backward.

Slowly moving down the trunk of the great pine-tree appeared an animal of enormous size. Had they not been expecting something of the kind neither could have told that this moving object was an animal: since at first sight neither a head nor limbs could be distinguished — only an immense shapeless mass of brown, shaggy hair.

The instant after a huge hairy limb was protruded below, and then another both terminating in broad un-gulated paws, that in succession griped the rough bark of the tree, causing it to rattle and scale off.

Singular as its shape was there was no mistaking the animal that was making this retrograde movement. It was Bruin himself, descending the tree buttocks down-ward!

CHAPTER XI.

HAND TO HAND.

ALEXIS and Ivan, as they started back, simultaneously screamed out a shout of warning to **Pouchskin**. Both, almost at the same instant, raised their guns, and fired into the buttocks of the bear.

Pouckskin had heard their cries, but not the preliminary "sniff" which the animal had uttered: he had been too eager in *listening inside of the cave*, to hear aught that was passing without. He heard their warning cry, however, and the reports of their guns; but not in time to get out of the way. Just as the shots were fired, he had half risen from his recumbent attitude; but the bear at that moment dropped down from the tree, and coming "*co-thump*" on the back of the old guardsman, once more flattened him out upon his face!

Perhaps it would have been as well for Pouchskin if he had quietly remained in that attitude: for the bear had already turned from him, and showed signs of an intention to retreat; but Pouchskin, deeming that he was in the worst position he could well be in, scrambled suddenly to his feet, and made a "grab" at his gun.

This show of fight on the part of his antagonist — and the belief, perhaps, that it was Pouchskin that had so

rudely tickled his posteriors — roused the fury of the
bear; and instead of exposing his hind quarters to a second assault, he charged mouth open upon the ex-guardsman. By this time, the latter had recovered his gun, and
promptly brought the piece to his shoulder; but, alas!
the gun snapped! The lock had been wetted in the snow-
trap. It was a flint lock, and the priming had got damped.

The failure only increased the fury of the animal; and
and a charge of swan-shot, which Ivan had the same instant fired from his second barrel, still further irritated
him.

Pouchskin drew his long-bladed knife. It was the only
weapon he could lay his hand upon, for the axe, which
might have served him better, had been left above on the
bank, where he had lopped the sapling.

He drew his knife, therefore, and prepared to defend
himself in a *hand to paw* struggle.

He might still have retreated, though not with a cer-
tainty of safety — for in the hurry of the moment the
bear had got on the bank above him; and had he turned
his back, the fierce quadruped might have overtaken, and
knocked him down at his will. Pouchskin thought it
better to face the bear, and receive his onslaught at
arm's length.

There was but one way in which he could have retreated, and that was backward down the slope. He
might make ground in that direction; and it occured
to him to do so, in order to get footing on a more level
surface.

The bear having paused a moment to bite the place
where the rifle bullet had stung him, gave Pouchskin time

to gain some ground backwards; but only a few paces—
since the whole affair did not occupy **a tenth** of the time
taken in describing it.

Just as Pouchskin had reached the bottom of the slope,
his angry assailant, with a terrific growl, rushed forth
from the smoke, and galloped directly towards him. When
about three feet distant from the hunter, Bruin reared
upon his hind legs, in the attitude of a prize-fighter!

Pouchskin was seen to lunge forward with his right
arm — the one which carried his knife, and, the moment
after, both man and beast appeared closed together, "in
grips."

In this fashion they went waltzing over the snow, the
spray of which rose in a cloud around them; and for
a while they were seen only as one dark, upright form,
in confused and violent motion!

Ivan was uttering cries of fear — fear for the safety
of his dearly-loved Pouchskin; while Alexis, more cool,
was rapidly reloading his rifle, — knowing that the surest
means of saving the life of their faithful attendant was
to encompass the death of the bear.

It was a moment of real peril for Pouchskin. The
bear was one of the largest and fiercest he had ever encountered; and, perhaps, had he examined the brute more
minutely before the conflict commenced, he would have
thought twice before facing him. But the smoke from
the guns was still over and around the spot, hanging upon the damp air. Up to the time when Pouchskin resolved to make stand, he had not yet had a clear view of
his shaggy antagonist. When at length he perceived the
formidable proportions of the animal, it was too late to
retreat; and the struggle began as described.

In brief time Alexis — who at loading was quick as a tirailleur — had recharged his piece, and was now hastening up to the rescue.

Without going quite close he dared not fire: for, in the way that man and bear were dancing about, there would be as much danger of killing the one as the other.

All at once, however, they appeared to separate. Pouchskin had torn himself out of the bear's clutches, and, evidently disinclined to a renewal of the embrace, was retreating backward, over the snow, still hotly pursued by the animal.

At this moment Alexis would have fired; but, unfortunately, the direction in which Pouchskin was going kept his body nearly in a line with that of the animal; and Alexis could not fire without danger of hitting him.

The chase led across the ravine, and of course over the bed of snow. The pursued was doing his best to escape. But the pursuer had the advantage — for while the man was breaking through at every step, the broad-pawed quadruped glided over the frozen crust without sinking an inch.

Pouchskin had got a little the start, but his pursuer was fast gaining upon him. Once or twice, indeed, the bear was close enough to touch Pouchskin's skirts with his extended snout; but the necessity of rearing up, before making a stroke with his paw, required him to get still nearer, and Bruin knew that.

He had, however, got near enough even for this; and had risen on his hind feet, with the intention of clawing down his victim. Ivan and Alexis simultaneously uttered a cry of dismay; but before the dangerous stroke

E

could descend, he for whom it was intended had sunk out
of sight!

At first, the young hunters believed the blow had been
struck, and that Pouchskin had fallen prostrate under it.
They saw the bear spring forward as if to cover the
fallen man; but the next moment their terror was mingled with astonishment on seeing, or rather *not* seeing,
either man or bear: both had suddenly disappeared!

CHAPTER XII.

A MYSTERIOUS DISAPPEARANCE.

THE sudden disappearance of both man and bear would no doubt have sadly perplexed our young hunters, had it not been for Pouchskin's previous adventure. With that still fresh in their memory, they were at no loss to comprehend what had occurred. While eagerly endeavoring to escape from his antagonist, Pouchskin had, no doubt, forgotten the dangerous snow-bridge; and, just as before, he had broken through it.

This time, however, it was no laughing matter. Pouchshin was no longer playing a solitary Jack-in-the-box, but, in all likelihood, he was under the huge body of the savage monster, in the act of being torn to pieces by his teeth, or perhaps drowned in the *subnivean* stream. Whether the bear had sprung voluntarily after him, or, in the impetus of charging, had been himself precipitated into the snow chasm without the power of preventing it, could not for the moment be known. The young hunters suspected that the bear had fallen in rather against his will; for certainly he had been seen to go down in rather an awkward and blundering manner, his hind legs pitching upwards as he broke through.

Whether the plunge had been voluntary or against his

will could matter but little. He must be now upon top
of the ex-guardsman ; and, knowing **the** implacable fury
of these animals when roused to resentment, his young
masters had no other idea but that their attendant would
be either drowned or torn to pieces.

As a last hope, however, Alexis rushed on over the
snow, holding his rifle before him, and prepared to fire
its contents into the bear the moment he should get sight
of the animal.

As he advanced, he could **hear a** plunging and splashing of **water,** with other noises, — **as** the snorting and
growling of **the** bear, and the crashing **of** frozen snow,
all mixed up in confusion of sounds. Concluding that
these noises were caused by the struggle still going **on**
between the man and the bear, he hurried **forward.**
Strange ! there came no voice from Pouchskin !

When within about three paces of the broken edge, an
object came under his eyes, that caused him to halt in his
track. That object **was the** snout of the bear, that was
projected upward above the surface of the snow. The
eyes of the animal were not visible, **nor** any other part
of it, except the aforesaid snout, and about six inches
of the muzzle.

The thought instantly occurred to Alexis, that the bear
had reared upon his hind feet, and was endeavoring to
clamber out ; and this was true enough, for **the** instant
after, he was seen to spring perpendicularly upward, until
his whole head and part of his neck became visible.
Only for an instant, however ; for Bruin, who now appeared to be playing Jack-in-the-box, sank once more out
of sight, snout and all.

The young hunter was just regretting that he had not taken a snap shot at the animal's head; but before ten seconds of time had elapsed, the snout was again popped up by the edge of the hole. In all probability the bear would make a second attempt to spring out.

Alexis was therefore waiting till the whole head should show itself; but quick as a flash of lightning, it occurred to him that the brute might at the second effort succeed in reaching the surface of the snow, and then he would himself be in danger. To avoid this contingency, he resolved to fire at once; not at the snout, for, although he could not have failed to send his bullet through it, he knew that that would not kill the bear, but only render him more desperately furious, if such a thing had been possible.

It was the bear's skull he meant to take aim at. From the position of the animal's snout, of course he could tell exactly where the head must be, though he could not see it.

Had Alexis been an unskilled marksman, he would have stood his ground; and, guessing the position of the bear's head, would have fired at it through the snow. But he did not act in this manner. He had scientific knowledge sufficient to tell him that his bullet, sent in a slanting direction, might glance off the frozen crust, and miss the mark altogether. To insure its direction, therefore, he instantly glided two steps forward, poked the barrel of his piece through the snow, until the muzzle almost touched the head of the bear, — and then fired!

For some seconds he saw nothing. The smoke of the

gunpowder, as well as the snow-dust blown up before the muzzle of the gun, formed a dense cloud over the spot. But though Alexis could not see the effect of his shot, he could tell by what he heard that his bullet had done good work. A loud "swattering" at the bottom of the hole proclaimed that the bear was struggling in the water; while his piteous whines and faint grunting told that his fierce strength was fast passing away.

As soon as the smoke had cleared off, Alexis upon his knees crept forward to the edge, and looked over it. There was blood upon the snow; **the side** against which the bear had stood was crimsoned with streams of it; and below, in the water, among the clumps of broken snow-crust, appeared a dark brown mass, which Alexis knew to be the body of the animal.

It was still in motion; but as it was in a prostrate attitude, and making only feeble efforts, the young hunter knew that the life was nearly out of it.

It was not this that was now causing him to look down with such an anxious and troubled countenance. It was his apprehensions for Pouchskin. Where was he? At the bottom of the crater-like pit Alexis could see the body of the animal, but nothing of a man, — neither arms, legs, nor body. Could he be under the bear, concealed by the shaggy hair? Was he hidden under the black water that filled the bottom of the ravine? — or, horrible thought! was he dead, and had his body been carried off by the current that rushed rapidly under the snow?

This was **not** improbable, for Alexis could see that there was a sort of arched tunnel between the snow and the water, quite large enough to have admitted the body of a man!

In agony he cried out, calling Pouchskin by name. He was repeating his despairing invocation, when all at once a loud laugh echoed in his ears, uttered close behind him. In the laughter he recognized the voice of Ivan.

Alexis suddenly leaped to his feet, wondering what on earth could be the cause of this ill-timed merriment. He turned towards Ivan with the intention of chiding him; but at that moment an object fell under his eye, that hindered him from carrying his intention into effect. On the contrary, the sight he saw caused him such joy, that he could not restrain himself from joining Ivan in his laughter. No wonder. The sight was odd enough to have drawn a smile from a dying man. A spectacle more ludicrous could scarce have been conceived.

A little further down the ravine, and about ten paces from where the boys were standing, an object was seen protruding above the snow. It was about ten inches in vertical diameter, something less horizontally, and of a roundish or oval shape. In color it was almost white as the snow itself: for, indeed, it was sprinkled over with this material out of the bosom of which it had just emerged. A stranger coming upon the ground might have been sorely puzzled to make out what it was; but not so Ivan, who, on first beholding it, as it popped upward through the frozen crust, recognized it as the head of Pouchskin. Alexis also made it out at the first glance; and it was the comic twinkle of Pouchskin's eyes — denoting that no great damage had happened to him — that led Alexis to join his brother in the laughter.

Their merriment, however, was of short continuance —
only an involuntary burst, for a moment's reflection told
them that Pouchskin, although they saw him alive, might
nevertheless have sustained some serious injury ; and both
at the thought hastened up towards the head.

On getting close to it, however, Ivan was unable to
control himself, and once more gave way to a fit of involuntary laughter. The head of the old guardsman,
standing up like a sphinx above the frozen surface, —
his grizzled hair powdered all over with snow like the
poll of some grand flunkey, — his long moustache loaded
with it, — his **eyes** sparkling and twinkling, and his
features set in a serio-comic expression, — all combined
to form a picture that it was difficult to contemplate with
seriousness.

Alexis, however, anxious to ascertain **as to whether**
Pouchskin had received any dangerous wound, did **not**
this time join in his brother's mirth ; and, as soon as they
came near enough, his inquiries were directed to that
end.

" Only scratched a bit, masters!" answered the old
guardsman, — " only scratched a bit — nothing much ;
but the bear — the bear ! where has the brute gone ? "

" To his long home," answered Alexis ; " you need be
under no further apprehension about him. I think your
knife must have wellnigh settled his account, for he was
unable to get out of the hole again ; but, fortunately, I
have finished him with a bullet, and it only remains for
us to haul his carcass up and take the skin off it. First,
however, let us endeavor to extricate you, my good
Pouchskin ; and then you can tell us by what means you

have managed to make an escape, that certainly appears miraculous !"

So saying, Alexis, assisted by **Ivan,** commenced digging away the hard crust that surrounded the neck **of** Pouchskin ; and kept on at it, until they had uncovered his shoulders. Then seizing him by the arms — one **on** each side — **they drew** him up, till his feet once **more** rested on the surface of the snow.

4

CHAPTER XIII.

A SUBNIVEAN ESCAPE.

POUCHSKIN proceeded to describe the manner of his escape — his young masters listening to him with great interest — although they already guessed pretty nearly how it had been accomplished. Still there were some points not so clear to them, which the old guardsman detailed.

In the first place, he had retreated from the bear, not because he believed himself vanquished, but because he had lost his knife. Its handle, wet with blood, had slipped from his grasp ; and he could not tell what had become of it ! Finding himself unarmed, of course his next thought was to get out of Bruin's way, for what could an unarmed man do in the embrace of a bear — and such a bear ?

He then turned and ran; but he had quite forgotten the dangerous character of the snow-bed — the bridge that had refused to carry him before ; though, indeed, over it was the only direction he could have taken. Had he attempted to run to the right or left, his course must have been up-hill ; and the bear would have been certain to overhaul him in a couple of leaps. After all, he had taken the proper direction; and, as it proved in the end, his breaking through was the most fortunate accident that could possibly have happened to him. Had it not chanced so, he would, in all probability, have fallen into the clutches of the bear, and been torn to shreds by the infuriated animal.

Well, on touching bottom, he felt the water among his feet, and just then remembered how it had been before. He remembered the hollow arch-way under the snow, and, seeing the bear above, and in the act of being precipitated on top of him, he suddenly ducked his head, and pushed himself into the tunnel. He could feel the bear falling upon him behind, and the weight of the animal's body, as it was precipitated downwards, forced him still further under the snow-bridge.

Once in, he continued on down the stream, working both with head and arms, and clearing a space that would allow his body to pass. The soft snow was easily pressed out of the way; and, after going as far as he deemed necessary, he turned to the right, and worked his way upward to the surface.

It was while he was thus engaged that Alexis had been squaring accounts with the bear. The fierce creature had not followed Pouchskin under the snow. In all probability, his sudden "souse" into the water had astonished Bruin himself;—from that moment all his thoughts were to provide for his own safety, and, with this intention, he was endeavoring to get back to the surface of the snowdrift, when Alexis first caught sight of his snout.

At the moment that Alexis fired the final shot, or just a little after it, Pouchskin had popped up his head through

the congealed crust of the snow, and elicited from Ivan
those peals of laughter that had so much astonished his
brother.

Pouchskin, however, had not come unscathed out of
the "scrimmage." On examining the old guardsman, it
was found that the bear had clawed him severely; and a
piece of skin, of several inches square was peeled from
his left shoulder. The flesh, too, was rather badly lacer-
ated.

Alexis was not without some surgical skill; and, without suffering a moment to be lost, he dressed the wound
in the best manner possible under the circumstances. A
clean handkerchief, which Ivan chanced to have, served
as a covering for the scar; and this being tied on securely, with a strip torn from the sleeve of Pouchskin's
own shirt, left the wounded guardsman in a condition to
recover, as soon as it might please nature to permit.
Nothing more could have been done by the most "skilful practitioner."

Their next business was to look after the bear. On
going back to the hole, and gazing into it, the animal, as
Alexis had anticipated, was quite dead; and the water,
partially dammed up by the huge carcass, was flowing
over it.

Ivan, who had hitherto done least of all to secure the
prize, now became the most active of the three; and,
leaping down upon the body of the great brute, he looped
the rope around one of its hind legs, and then stood on
one side to help the rest in raising it upward.

Alexis and Pouchskin commenced hauling on the other
end of the rope, and the vast mass slowly ascended up-

ward, Ivan pushing from below, and guiding it past the inequalities of the snow. It would have been a different sort of a task, to have hauled Bruin out of such a hole three months earlier in the season; that is, about the time he had lain down for his winter *siesta*. Then he would have turned six or seven hundred pounds upon the scales, whereas at this time he was not more than half the weight. His skin, however, was in just as good condition as if he had been fat; and it was this, and not his carcass, that our hunters cared for.

After some tough pulling, accompanied by a good deal of shouting from Ivan at the bottom of the hole, the huge carcass was dragged forth, and lay at full length along the frozen snow. It was still necessary to raise it to the branch of a tree, in order that it might be skinned in a proper manner. This, however, could be easily accomplished by means of the rope.

Up to this time Pouchskin had been puzzled about the loss of his knife. Everywhere he looked for it; but it was nowhere to be found. All the surface over which he had danced with the bear was carefully examined, and the snow scraped up to the depth of several inches. There was the blood of the bear, and some of Pouchskin's own too, but no knife! Could it have got into the water? No. Pouchskin declared that he had dropped it near the edge of the snow-bed: for this accident, as already stated, had been the cause of his retreat from the conflict.

It was only when the great carcass was being hauled up to the branch, that the lost knife made its appearance. Then, to the astonishment of the young hunters, as well

as to Pouchskin himself, the knife was seen sticking in
the shoulder of the bear! There it had been when the
haft slipped from his hands, and there had it remained.
No doubt that stab would have given the bear his death-
blow; but still more fatal had been the bullet from the
rifle of Alexis, which had passed through Bruin's brain,
crushing his skull like a shell!

The skinning of the animal was accomplished with
great care; for the coat **was** one of the finest, and the
boys knew with what interest it would be regarded on its
arrival at the palace Grodonoff. They spared no pains,
therefore, in removing it from the carcass; and after the
work was finished, it was neatly folded up, tied with the
rope, and placed like a knapsack on Pouchskin's shoulders.

Of the carcass they took no heed; but leaving it to
the wolves, the gluttons, or any other carnivorous creatures that might chance to stray that way, they turned
back up the ravine; and, striking off on a path that led
towards the tent of the Laplander, reached their smoky
quarters in good time for dinner.

CHAPTER XIV.

RINGING THE BEAR.

THE bear thus killed was the true *ursus arctos*, or brown bear — the latter name being given to him from the color of his fur, which, in ninety-nine cases out of every hundred, is a uniform brown. The name, however, is not appropriate, since there other brown bears belonging to very different species.

Having secured his robe, as we have seen, the next care of our hunters was to obtain a skin from the body of his black brother. They were well aware that this would not be so easy of accomplishment, from the simple fact, that the *ursus niger*, or "European black bear," is one of the rarest of animals — indeed, so few of them are obtained, that out of a thousand skins of the European bear that pass through the hands of the furriers, not more than two or three will be found to be of the black variety.

It is true that they were just in the country where they would be most likely to fall in with one; for it is only in the northern zone of Europe (and Asia also) where the black ones are found. This variety is not encountered in the southern ranges of mountains in the Alps, Pyrenees, and Carpathians. Whether this black

bear is a distinct species was not a question with them.
They knew that by most naturalists he is recognized as a
variety — by some a permanent one. It was, therefore,
certainly included in the conditions of their father's letter ; and a skin must be procured *coûte qui coûte.* This
done, they would have no further business in Lapland,
but might proceed at once to the Pyrenees.

It was not necessary to procure skins of the gray or
silver bear, nor that with the white ring round its neck —
known as the ringed or collard bear. As Alexis had
said, it is acknowledged by all who know the *ursus arctos*
in his native haunts, that these are mere accidental varieties. The true " collared bear " (*ursus collaris*) is not
found in Lapland, — only in northern Asia and Kamschatka, and it is he that is known as the " Siberian
bear." The boys therefore were not "bound" by their
covenant to procure these varieties ; but for all that, they
were gratified at going beyond the strict letter of their
agreement, which good luck enabled them to do ; for,
while scouring the country in search of the *ursus niger*,
they chanced upon another brown bear, a female, with
three cubs, one of which was brown, like the mother ;
the second had the white ring round its neck, and the
third was as gray as a little badger ! All **four were**
taken ; and the young hunters not only had the gratification of being able to send the different varieties of skins to
their father's museum, but an additional satisfaction was
afforded **to** Alexis, the naturalist, by this grand family
capture. It proved incontestably, what he already suspected, and what, moreover, the native peasants and hunters had told him, that the " silver " and "ringed " bears
were identical with **the** *ursus arctos.*

Notwithstanding their joy at the capture of the old she and her particolored pets, they were yet very anxious about the black bear. They had hunted all the forests and mountains for miles around, and had even succeeded in killing several other specimens of " Brownie," but no " Blackie " was to be met with.

It had now got known among the native hunters what they were in search of; and, as they had offered a liberal reward to any one who could guide them to the haunt or den of a real black bear, it was not unlikely they should soon hear of one.

In this expectation they were not deceived. About a week after the offer had been proclaimed, a Finnish peasant (one of the Quäns, as they are called) made his appearance at their head-quarters, and announced that he had " ringed " a black bear. It was welcome tidings; and the young Russians at once proceded to the indicated place.

It may be necessary to explain what the man meant when he told them he had " ringed " the bear; since that is a phrase of specific meaning throughout the countries of Scandinavia. In these countries, when the track of a bear is observed in the snow, it is followed up by the person who has discovered it, with the intention of " ringing " the animal — that is, ascertaining as near as may be, the locality in which it may have halted from its rambles, and lain down to rest. Of course, if the person thus trailing the bear be a hunter — or if it be a party of hunters actually engaged in the chase, they will keep on until they have found the bear in his den. But in nine cases out of ten, bears are not pursued in this fashion.

4 *   F

Generally, their haunt — whether temporary or otherwise — has been ascertained beforehand, by some shepherd or wood-cutter, and a party of hunters then proceeds to the spot, and makes a surround of the animal before rousing him from his lair.

This "surround," however, has nothing whatever to do with the "ringing" of the bear, which is an operation of a different character, and is performed by the party who has first chanced upon the tracks. The mode of proceeding is simply to follow the trail, or *spär*, of the bear as silently as possible — until the tracker has reason to believe that the animal is not far off. This he discovers by observing that the *spär* no longer trends in a direct line, but doubles about in zigzags, and backward turnings upon itself; for when a bear intends to lie down, it is his habit to quarter the ground in every direction, precisely as does the hare before squatting in her form. Many other animals observe a similar caution before going to rest.

The bear-tracker having reached this point, then leaves the track altogether, and makes a circuit round that part of the forest within which he suspects Bruin to have couched himself. This circuit is of greater or less diameter, according to circumstances — depending on the season of the year, nature of the ground, and a variety of other considerations. While going round this circle, if it should be seen that the track of the bear leads beyond it, then that "ring" is given up, and another commenced further forward. If, on the other hand, the the tracker gets round to the place whence he first started, without again coming upon the *spär*, he concludes that the bear must be lying somewhere within the circumference which he has traced, and will there be found. This, then, is termed " ringing" the bear.

You may wonder why the man does not follow up the *spär* until he actually reaches the den or lair of the animal. That is easily explained. The tracker is not always a bear-hunter, and even if he were, it would not be prudent for him to approach a bear without assistants, who, by surrounding the animal, should cut off its retreat. Were he to go forward direct to the bear's hiding-place, Bruin would, in all probability, discover him before he could approach within shot; and, making a bolt, might carry him a chase of ten or twelve miles before stopping. The brown bear often does so.

The tracker, having ascertained the circle within which the animal has made its temporary resting-place, next proceeds to warn the hunters of his village or settlement; and then a large party go out for the destruction of the common enemy. They deploy around the ring, and closing inward, are pretty sure to find the bear either asleep in his den, or just starting out of it, and trying to get off.

The " ring" will usually keep for several days — sometimes for weeks — for the bear, especially in winter time will remain in the vicinity of his lair for long spells at a time. Frequently several days will elapse before any hunters arrive on the ground ; but if the bear should have strayed off in the mean time, his tracks in the snow will still enable them to follow and find him. If, however, fresh snow should have fallen, after the bear has made his exit from the marked circle, then, of course, the

search will prove a blank, and Bruin make his escape —
at least out of that " ring."

One of the most singular features of this custom is,
that he who has succeeded in " ringing " a bear, is regarded as the lawful proprietor of the animal — or rather
of the " ring " and can dispose of his right to any hunt-
ing-party he pleases. Of course he cannot guarantee the
killing of the bear : that is left to the skill of the hunters,
who must take their chance. **The tracker** only answers
for a bear being found within a prescribed circle, of which
he gives proof by pointing out the *spär*. With such
conditions, established by long and well-observed custom,
it will easily be believed that the wood-cutters and other
peasants make a market by ringing bears, frequently
disposing of the " ring " to the more ardent hunters for
a very considerable price! It was just with this view
that the Finnish peasant had put himself in communica-
tion with our young Russians ; and as the bounty they
had already offered far exceeded the usual purchase-
money in such cases, **the** Quän at once closed with their
offer, and conducted them to the " ring."

CHAPTER XV.

OLD NALLE.

WHILE proceeding towards the ground where they expected to find the bear, their guide informed them that he had not only ringed the animal, but actually knew the den in which it was lying. This was still better: it would not only save them a search, but enable them to encompass the beast on all sides, and cut off his retreat — should he attempt to bolt before they could get near.

On approaching the place, therefore, Pouchskin proposed that the three should separate, and, after having deployed into a circle, proceed inward from different directions.

But the guide opposed this suggestion — saying, with a significant smile, that there was no need of such precautions, as he would answer for the bear not leaving his den, until they had all got up as near as they might wish to be.

The hunters wondered at this confidence on the part of their guide, but in a few minutes' time they had an explanation of it. Going up to a sort of cliff that formed the side of a little stony knoll, the Quän pointed to a hole in the rocks, saying, as he did so: —

"Old *nalle* is in there."

Now "nalle" is the nickname of the bear throughout the Scandinavian countries, and our Russian hunters knew this well enough; but that a bear could be inside the little hole to which their guide had pointed appeared utterly incredible, and Ivan and Alexis burst into a loud laugh, while Pouchskin was rather inclined to show a little anger about the matter.

The hole which the Quän had pointed out was a crevice between two great boulders of rock. It was about a yard above the ground upon which they stood; and was certainly not more than six or eight inches in diameter. All round the orifice the rocks were thickly coated with ice; and from the top of the cliff on both sides huge icicles projected downwards, until their tips touched the earth, looking like enormous trunks of elephants, or such as even mammoths might have carried. One of these immense icicles was directly in front of the aperture; while on the ground just below its point stood up a huge mass of an irregular conical shape, the convex surface of which was coated with snow that had lately fallen.

The first impression of the hunters was, that they had been deceived by the cunning Quän. Pouchskin declared that they would not stand being tricked; and at once demanded back the ten rix-dollars which his young masters had paid for the "ring" of the bear.

"It was all nonsense," he said; "even if there was a cave, no bear could be inside, for the simple reason that none, even the smallest, could possibly have squeezed his carcass through a hole like that; — a cat could hardly have crept into such an aperture! Besides, where were the tracks of the bear? There were none to be seen —

neither by the mouth of the hole, nor in the snow out-side."

There were old tracks of the peasant himself and of a dog, but not of a bear.

"It is a decided take-in," grumbled Pouchskin.

"Patience, master!" said the Quän. "There is a bear inside for all that; and I'll prove it, or else return you your money. See my little dog! he'll tell you old *nalle* is there. It was he that told me."

As the Quän said this he let slip a diminutive cur, which he had hitherto held in the leash. The animal, on being set free, rushed up to the hole, and commenced scratching at the ice, and barking in the most furious and excited manner. It certainly proved there was some living creature inside; but how could the Quän tell it was a bear; and, above all, a black bear!

He was interrogated on this point.

"By it," replied the peasant, taking from his pouch a tuft of long black fur, which was evidently that of a bear; "that is how I know that old *nalle*'s in the cave, and the color of the hair tells me that it's *black nalle* who's inside."

"But how came you by that?" inquired all three in a breath, as the man held the tuft before their eyes.

"Well, masters!" answered the Quän, "you see some jaggy points on the rock, at the top of the hole, there. I found it sticking there, where the bear must have left it, as he was squeezing himself into his cave — that's how it was."

"But surely," said Alexis, "you don't mean to assert that a bear could pass through such a hole as that? Why, a badger could n't get in there, my man!.

"Not *now*," said the Quän, "I admit; it 's three months since he went in. The hole was bigger then."

"Bigger *then*?"

"Certainly, masters! the heap you see below is only ice. It 's the drip of that great icicle that has frozen up as it fell, and if it were not there you 'd see a place big enough for a bear to get in. Ah! sirs! he 's there, I can assure you."

"Why, he could n't get out of himself?"

"That is very true," replied the peasant; he 'd be safe enough there till a good bit on in the spring. If we had n't found him, he would have been obliged to stay in his cave till the sun had thawed that great heap out of his way. It often happens so with the bears in these parts," added the Quän, without seeming to think **there** was anything unusual about the circumstance.

What the man said was literally true. The bear had gone into this cleft or cave to take his winter nap, and during the long weeks, while he was thus hybernating, the water, of rain and melting snow, dripping from the top of the cliff, **had** formed enormous stalactites of ice, with stalagmites as well: since it was one of the latter that had closed up the entrance to the den, and fairly shut him up in his own house!

Not only does this curious accident often occur to Scandinavian bears, but these animals, notwithstanding their proverbial sagacity, frequently become their own jailers. They have a habit of collecting large quantities of moss and grass **in** front of their caves, which they place right in the aperture; and not inside as a bed to lie upon. Why they do so is not clearly understood. The Scandinavian hunters allege that it is for the purpose of sheltering them from the cold wind, that would otherwise blow up into their chamber; and in the absence of any better explanation this has been generally adopted. The heap soon gets saturated by rain and melting snow, and congeals into a solid mass, so hard that it requires to be cut with an axe before it can be got out of the way; and the bear himself is totally incapable of removing it. The consequence is that it often shuts up the entrance to his winter chamber; and Bruin, on awakening from his sleep, finds himself caught in a trap of his own construction. He has then no other resource but to remain inside till the spring heats have thawed the mass, so that he can tear it to pieces with his claws, and thus effect an exit. On such occasions, he issues forth in a state of extreme weakness and emaciation. Not unfrequently he is altogether unable to clear away the obstacle, and perishes in his den.

On hearing these explanations from the Quän, who appeared to be well acquainted with Bruin's habits, the young hunters were satisfied that a bear was really in the cave. Indeed, they were not long upon the spot, till they had still more satisfactory evidence of this fact; for they could hear the "sniffing" of the animal, with an occasional querulous growl, as if uttered in answer to the barking of the dog. Beyond doubt, there was a bear inside.

How was he to be got out? That now became the important question.

CHAPTER XVI.

THE STAKED ENCLOSURE.

THEY waited, for a time, in hopes that he might show his snout at the little aperture, and all three stood watching it, with guns cocked and ready. A good while passed, however, and, as no snout made its appearance, they came to the conclusion that the bear was not to be caught in that simple way. By the snorting growl they could tell that he was at no great distance from the entrance, and they thought a pole might reach him. They tried this, but found that it could be inserted only in a diagonal direction; and although Pouchskin poked in the pole, and bent it round like a rattan, he could not touch wool anywhere; while the bear, though he gave tongue now and then, still kept his place at the further end of the cave.

No other plan offered, except to cut away the icy mass, and set open the mouth of the cavity. If this were done, would Bruin be then likely to come forth? The Quän was confident he would; alleging as his reason, that, in consequence of the spell of warm weather there had been, the bear must have fully shaken off his winter drowsiness, and would no doubt have been abroad long ago, but for the ice preventing his egress from the den. As soon as that should be removed, he would be pretty sure to sally out — for hunger, said the peasant, will bring him forth, if not just at the moment, certainly within an hour or so. At the worst they could wait awhile. Moreover, were the ice removed, they might be able to reach him with a pole; and that would be certain to put him in such a rage as would at once tempt him to make a *sortie*.

With this idea, Pouchskin seized his axe, knocked the great icicle into "smithereens," and was about going to work upon the huge *stalagmite* that blocked up the entrance, when he was interrupted by the Quän.

"With your leave, master!" said the latter, as he laid his hand upon Pouchskin's arm to restrain him. "Not so fast, if you please."

"Why?" asked the ex-guardsman, "don't you intend to unearth the brute?"

"Yes, master," replied the Quän; but something must be done first. This is a black bear, you must know."

"Well, and what of a black one more than any other?" demanded Pouchskin, somewhat surprised, for in the forests of Russia, where he had hunted bears, there were no black ones.

"Don't you know," said the Finn, "that black nalle is always bigger and fiercer than his brown brother? Besides, just at this time he will be so savage with hunger that he would eat one of us up the moment he got out. If that ice was away, I should n't like to stand here. Take your time, master! I think I can show you a better plan; at all events it is a safer one. It's a way we practise here — when we are sure that a bear is asleep, and won't interrupt us while we're making ready for him."

"Oh, well," replied Pouchskin, "I'm agreeable to anything you propose. I'm not particularly desirous of risking another wrestle — not I — I had enough of that the other day." And as the old guardsman made the remark, he gave a significant shrug of his shoulders, the wounds upon which not being yet quite cicatrized, feelingly reminded him of the rough handling he had received.

"Well then," said the Quän, "if you will help me to cut some strong stakes, I shall show you a plan by which you may knock old *nalle* upon the skull without danger to any of us, or send your bullets through his brain, if you like better to kill him in that way."

All, of course, agreed to the Quän's proposal; **for if** the black bear was, as he represented him, fiercer than his brown brethren, it would be no pleasant prospect to have him loose among them; and in case of their not being able to shoot him **dead** on the spot as he rushed out, they might not only be in danger of getting mauled, but in danger of what they dreaded almost as much — losing him altogether. He might get off into the forest; and as there were tracts along the hill-sides, now quite clear of snow, he might steal away from them beyond recovery. This would be a disappointment of no ordinary kind. In fact, it might be the means of keeping them for weeks, or perhaps months, from proceeding on their journey: since it might be weeks or months before they should fall in with another chance of obtaining a black bear-skin; and until that was procured they could **not turn** their faces towards Spain.

With such a prospect then, they were only too ready
to agree to any conditions by which the bear might be
safely secured. The Quän was not long in disclosing his
plan; and as soon as he had communicated it, all three
set to work to aid him in its execution.

A number of stout stakes were cut — each about six
feet in length, and pointed at one end. These were
driven into the earth around the outer edge of the icy
mass, in a sort of semicircular row; and so as to enclose
a small space in front of the aperture. To hold the
stakes all the more firmly, large stones were piled up
against them, and the uprights themselves were closely
wattled together by the broad flat branches of the spruce
pines that grew near. In this way was constructed a
fence that a cat could not have crawled through, much
less a bear. One aperture only was left in it, and that
was directly in front — a hole at about the height of a
man's knee from the ground, and just big enough to
admit the head of a bear — for that was the purpose for
which it was intended.

The next thing done was to roof the whole of this
stockade enclosure; and that was accomplished by resting long poles horizontally over it, tying them at the ends
to the tops of the uprights, and then covering them
thickly with *granris* (the spray lopped from the branches
of the evergreen pines).

It now only remained to get the ice out of the way,
and allow the bear to come forth. That would not have
been so easy of accomplishment, had it not been already
partially removed. Before closing up the top, Pouchskin, directed by the Finnish peasant, had cut away most

of the mass, leaving only a shell; which, although filling up the entrance as before, could be easily beaten down, or driven in from the outside of the enclosure.

During the time that the ex-guardsman had been sapping away the ice, he had been keeping a sharp lookout. He was admonished to do this by certain noises that, now and then, came rumbling out of the cave; and not very certain that he was in perfect safety, he had been under some apprehension. The bear, by throwing all his weight against the reduced mass of ice, *might* break his way out; and as by the constant chiselling the wall grew weaker and thinner, Pouchskin's fears increased in proportion. He was only too happy, when, having picked the congealed mass to what was thought a sufficient thinness, he desisted from his work, and crept out of the enclosure, through the space that had been kept open for him.

This was now fenced up as securely as the rest; and it only remained to knock away the icy barricade, and tempt Bruin to come forth.

The icy wall could be broken in by means of a long boar-spear with which the Finnish peasant had provided himself. It was headed with a heavy piece of iron, edged and tipped with the best Swedish steel, and this being jobbed against the ice, and kept constantly at work, soon splintered the shell into pieces.

As soon as the Quän saw that he had opened a hole large enough to pass the body of the bear, he drew back his spear, telling the hunter to look out.

During the operation, all three had kept watch through crevices in the stockade-wall, holding their guns pointed towards the aperture, and ready to give the bear a volley **the moment** he should show his snout

CHAPTER XVII.

THE SPITTING DEVIL.

To their disappointment, however, the bear refused to exhibit even as much as the tip of his nose, not only while his door was being opened, but afterwards; and they began to think that he might not come forth after all.

The Quän assured them that he would be certain to do so in time; but perhaps not for a few hours, till after they should have remained quiet a while, and old *nalle* should fancy they had gone away.

"He has been a long time without his breakfast," added the Quän, "and his stomach will now be talking loudly to him; that will tell him to steal out for something to eat. No fear of it, masters!"

"But for what purpose have you designed this hole?" inquired Ivan, pointing to the little aperture that had been left in the wattled enclosure.

"O," replied the peasant, "that is how we kill bears sometimes; especially if we are not rich enough to have a gun. As soon as old *nalle* rushes out from his den, the first thing he does is to run all round, looking for a chance to break through the fence. Of course he finds the hole, and pokes his head through it. One of us stands outside, as you see me now, with a hatchet ready; and we would be clumsy, indeed, if we could not cleave in his skull, or give him such a crack upon it, as would turn him back downwards. You shall see how the bear will rush to this hole the moment he comes out, and then, masters! you shall see!"

Here the Quän gave a significant shake to his hatchet, twirling it with the dexterity peculiar to his craft, for it so chanced that he **was** a wood-cutter by trade.

Our hunters, however, saw that this would never do. According to the conditions under which they travelled, the bear must be killed by one of themselves; and, therefore, after a little explanation, the Quän resigned his intention and stepped aside. His post, however, was supplied by the ex-guardsman, who, poising his ponderous axe, stood ready to deal a far heavier and deadlier blow than could be given by any wood-cutter in Scandinavia. Alexis took charge of Pouchskin's gun, determined to fire it as soon as he had discharged his own rifle; and as Ivan had one barrel loaded with ball and the other with slugs, it was not likely, against such a formidable battery **as was** thus prepared **for him,** that Bruin could manage to live much longer.

It now became a question whether they should wait patiently till the bear came out, or whether they might not adopt some mode of tempting him forth, that would act upon him more rapidly than the cravings of his stomach.

There could be no harm in trying to reach and stir him up with a pole; and for this purpose the wood-cutter stepped aside to find one.

He very soon succeeded in procuring a long birch sapling—as long as an ordinary fishing-rod; and having cleared this of its spray, he inserted it into the cave. To the gratification of the party it was found long enough for the purpose; for by the muffled feel it could only be Bruin's fur that its point was buried in. It was just as far, however, as the pole would reach; and as it was a slender sapling without any stiffness in it, they were unable to do anything in the way of giving him a poke. No doubt, had the entrance to his den been wider, even the tickling of the pole would have caused him to "turn out;" for a bear, unless badly wounded, will not stand much badgering. It was possible, in this case, that Bruin suspected there was some trap set for him outside—indeed, the noises he had been listening to for more than an hour must have admonished him that all was not as it ought to be; and this perhaps rendered him more wary than was his wont. He might not yet be aware that his door was open; for the roofed enclosure still kept out the light as much as the *stalagmite* had done; and although he might have heard the icy mass giving way before the axe and spear, he might not understand all that. It was necessary, therefore, to coax him as far as the threshold —so that he might discover that the door of his chamber had been opened for him.

The tickling of the pole, however, proved of no service; for, although it drew from the huge brute a sniff or two, he still kept to his bed.

What was to be done? Must they retire, and wait patiently till the calls of hunger should urge him forth? The day was piercingly cold, and to remain there long would have been unpleasant enough. They might, indeed, have to stay by the cave all day and all night too : for the enclosure had been only slightly put up — merely to check the bear for a few minutes — and if they were to leave him all night to himself, he could easily tear down the stakes and get off.

They could not think of deserting the spot for an instant ; but to avoid a long vigil they set about considering some plan by which Bruin might be induced to come forth from his inaccessible retreat.

A thought occurred to Ivan, who was a quick conceiver — a plan which promised well — and that was to make a " spitting-devil," and send it up into the cave. It appeared a good idea — at all events, it would not be difficult to give it a trial. Gunpowder was not scarce with them — since Russian roubles were plenty; and Pouchskin, pouring out nearly a quarter of a pound into the palm of his broad hand, commenced spitting upon it and working the powder into a paste. Ivan, who directed this operation, was determined his plan should not fail by any stinginess in regard to the materials required for carrying it out.

After a short space of time the plastic fingers of Pouchskin had elaborated the powder paste into a roll as large as a regalia cigar ; and this being dried slightly near a fire — which they had long before kindled — was ready for the touch. To the old grenadier was intrusted the management of the miniature rocket ; and, while the young hunters once more stood to their guns, he proceeded to carry out the design.

Having thrust his head through the hole intended for

the bear, and his arm through another which he had made for himself, he held the "devil" at arm's length between his finger and thumb. The Quän now took a blazing faggot from the fire, and passing it between the wattles, ignited the *fusé* which the old grenadier had ingeniously placed in the "devil's" tail.

As soon as Pouchskin perceived that it was fairly on fire, with an adroit jerk he sent the little rocket up unto the cave, as far as he could throw it; and then jerking himself backward, he seized hold of his axe.

There was a moment of suspense; not long: for almost on the instant a brilliant light shone within the cave, accompanied with a sputtering and whizzing and cracking, as if half a dozen alarm-clocks had been set going at the same time! In the midst of this confusion of noises, and louder far than any, could be heard a number of sharp wild shrieks, and before the rocket had half burnt out, Bruin was seen bolting forth over the broken fragments of ice. Two shots were fired, almost simultaneously; but both failed to check his onward rush; and with a mighty force he came "bump" against the palisades, causing them to crash and swag as if they would give way. It was fortunate for the hunters that the stakes stood the shock: for such a set of teeth as that bear exhibited they had never before seen. A single stroke from those paws would have been enough to crack the thickest skull in creation.

Ivan gave him his second barrel — the one loaded with slugs, — but it only served to increase his fury; and now rearing up, and then going on all-fours, he kept rushing backward and forward through the enclosure, all the while uttering fierce growls.

Alexis, meanwhile, had dropped his rifle and taken up the fusil of Pouchskin. His place was at one side of the enclosure. He had already got the barrel through the wattles, and was endeavoring to level it upon the bear — seeking for a mortal part at which he might aim. The darkness, however, — for the roofed stockade rendered it dark, — combined with the quick movements of the animal, hindered him from getting a sight to his satisfaction. He knew the importance of making this a killing shot. Should the bear, **wounded** as he now was, retreat back into his den, there would be no chance what**ever** of getting him out again. Alexis thought of this; and therefore resolved not to fire at random, as he had done before. He knew that a full-grown bear, unless shot in the brain or heart, can accommodate a score of bullets without being much inconvenienced by them.

Knowing this, Alexis was biding his time, when all at once he perceived the bear make halt on the front side of the enclosure. **He now** aimed **at** the heart of the huge animal, but before **he could** pull trigger, a loud crash sounded in his ear, and Bruin was seen dropping to the ground, where he continued to lie, almost without giving **a** kick!

It **was the axe** of Pouchskin that had caused the crash, as its edge of steel descended upon the bear's cranium, smashing it in as if it had been an egg-shell. As the Quän predicted, the animal had imprudently poked his head through **the** aperture where Pouchskin was standing ready for him.

This, of course, finished the affair. It was only necessary to remove the palisades, sling the bear to a tree,

and then strip him of his much coveted skin. All this in
due time was accomplished; and with the robe once
more packed on the shoulders of Pouchskin, the hunters
returned to their head-quarters.

It proved — as the Quän had promised them — to be
a black bear; not that his fur was altogether black, as is
the case with the *ursus americanus* and the black bears
of India. On the contrary, the hair was brown near the
roots, and only black at the tips, which, however, gave it
the appearance of being black all over the surface; and
Alexis knew that this was the variety of bear they were
in search of.

Satisfied that they had obtained the skin of the *ursus
niger*, it only remained for our hunters to pack up their
travelling traps, bid adieu to the cold climate of Scandinavia, and start for the sunny south — for the far-famed
Pyrenees of Spain.

BRUIN.

CHAPTER XVIII.

THE PALOMBIERE.

IT is not intended to detail the many incidents that befell them on the way, the chitchat of steamboats, rail-
roads, and hotels. Their father cared not to hear of
these trifles; he could read enough of such delightful
stuff in the books of whole legions of travellers; and, as
they did not note anything of this kind in their journal,
we are left to suppose that they encountered the usual
pleasures and *désagrémens* which all travellers must experience on similar journeys. As money was no object,
they travelled with expedition — making only a short
stay in the great capitals through which they passed, in
order to have their passports *viséd*, and sometimes for
the purpose of using the great emperor's letter for the
replenishment of their exchequer. This magic document proved all-powerful everywhere they went; and as
they knew it would be so in all corners of the habitable
globe, they could rely upon it with perfect confidence.
Pouchskin's leathern bag was always well weighted with
the yellow metal, — and *specie*, whatever stamp it may
bear, is current all over the world.

Their journal merely mentions the route followed.
From their hunting-ground they returned down the Tor-

nea river, which, running due north and south, of course
did not compromise the terms of their covenant; neither
were the conditions infringed by their taking at any time
the back-track when engaged in the chase, for, as already
known, there was a specification in the baron's letter that
allowed of this deviation. All that was required of them
was that they should not recross a meridian when on their
actual route of travel.

A ship carried them from Tornea to Dantzic. Hence
they passed to Berlin, and on through Frankfort, Stut-
gard, and Strasbourg, to Paris. Paris, it is true, was a
little out of their way; but what Russian could travel
across Europe without paying a visit to Paris? Pouchskin cared little about it. The old grenadier had been
there before, — in 1815, — when he was far from being
welcome to the Parisians; and Alexis would rather have
gone by another and more direct route, that is, through
Switzerland; but the gay Ivan would not hear of such a
thing. To see Paris he was determined; and see it he
did; though what he or they did there is not mentioned
in the book of the chronicles of our young bear-hunters.

From Paris they travelled by rail, almost directly
south — though still slightly westward — to the celebrated baths of Bagnères. Here they found themselves
not only within sight, but actually among the *foot-hills* of
those mountains, for the tourist scarce second in interest
to the Alps themselves, but perhaps for the naturalist
even more interesting than these.

At Bagnères they made but a short stay, only long
enough to recruit their strength by bathing in its thermal
springs, and to witness a spectacle which is regarded as
the grand lion of the place — the *Palombière.*

As you, young reader, may not have heard of the *Pa-lombière*, and may be curious to know what it is, I give the account of it which I find recorded in the journal of Alexis.

About two miles from Bagnères rises a ridge of con-siderable elevation — running parallel with the general direction of the Pyrenees, of which it may be considered an outlying step, or "foot-hill" (*pied mont*). Along the crest of this hill stands a row of very tall trees, from which the branches have been carefully lopped, leaving only a little bunch at the top of each. On coming close to these trees — provided it be in the months of Septem-ber or October — you will observe a something between them that resembles a thin gauzy veil of a grayish color. On getting still nearer, you will perceive that this veil is a net — or rather a series of nets — extended from tree to tree, and filling up all the spaces between them, from the highest point to which the branches have been lopped down to within three feet of the ground.

Another singular object, or series of objects, will long ere this have attracted your attention. You will see standing, at certain intervals apart, and about thirty yards in front of the trees, a row of tall tapering sticks — so tall that their tops are fifty yards from the ground! They might remind you of the masts of a ship ; but that there are in each case two of them together, — the one stand-ing vertically, and the other bending over to it, with a slight curve. On this account you may be more struck with their resemblance to the "shears" seen in shipyards, by which the masts are "stepped" into their places. These masts, as we may call them, are not all of one stick

of wood, but of several pieces spliced together; and, notwithstanding their prodigious length — fifty yards, you
will remember — they are of no great thickness. In
fact, although the two are joined together at the top —
as we shall presently have occasion to show — when a
strong wind blows, both bend, and vibrate back and forward like an elastic trout-rod. At their bases they are
only five feet apart; and the curving one is intended to
act as a stay to the other. Both, as already stated, meet
at the top, and looking up you will see — while the sight ·
makes you dizzy — a little roundish object at the point
of the junction. It is a basket set there firmly, and just
big enough to hold the body of a man. If you look carefully you will see a man actually within it; but, to quote
Shakespeare's quaint simile, he will appear to your eyes
not half as gross as a beetle! In all likelihood he is not
a man, but only a boy; for it is boys who are selected
to perform this elevated and apparently dangerous service.

How did the boy get there? will probably be your
next question. By running your eye along the curved
pole, you will perceive a row of projecting pegs extending from bottom to top. They are quite two feet apart;
but had you been present while that youth was making
the ascent — which he did by the help of these pegs —
you would have seen him scramble up as rapidly, and
with as little concern, as a sailor would ascend the rat-
lines of a ship! It is his trade to do so, and practice has
made him as nimble as he is intrepid; but you, who are
unaccustomed to witness such tall gymnastics, cannot help
again recalling Shakespeare, and exclaiming, with the

5 *

great dramatic poet, " Fearful trade !"   Quite as fearful,
indeed, as the gathering of " samphire."

But what is this trade ?   What is all this contrivance
for — these nets and tall masts, with " crows'-nests " **at**
their tops ?   What are the boys doing up there ?   And
what are they about below — those men, women, and
children — a crowd composed of all ages and all sexes ?
What are they doing ?

Pigeon-catching.   That is what they are doing, or rather
what they are aiming **to do, as soon as** the opportunity
offers.   These people are simply pigeon-catchers.

What sort of pigeons ? **and where do they come from ?**
These questions must be answered.

To the first, then, the answer is, the common European
wild pigeon (*columba palumbis*).   It is well known in
England by the name of " wood-pigeon," and in France
it is called *ramier.*   In England the wood-pigeon is not
migratory.   In that country there is a much milder winter than is **experienced** in the same or **even** a more
southerly latitude **on the** Continent.   This enables the
pigeon to find **food** throughout **all the year,** and it therefore remains in England.   **In** continental countries —
France among the number — the severity of the winter
forces it southward ; and it annually migrates **into Af-**
rica — the supposed limit of its flight being the chain
of **the** Atlas mountains.   Of course the wood-pigeon is
only one of many birds that make this annual tour, taking, as the rest do, **a** " return ticket."

Now, the *ramiers* of France, in passing southward,
must ply their wings a little more strenuously to mount
over the snowy summits of the Pyrenees ; but they only

commence ascending to this higher elevation when near the mountains. The ridge at Bagnères chances to lie in the line of their flight — of course, not of all of them, but such as may be sweeping along in that particular meridian; and, passing between the tall trees already mentioned, they get caught in the meshes of the nets. The moment they strike these — several of them coming "but" against one at the same instant, — a trigger is pulled by the men — who are below concealed under screens — and this trigger, acting on a string, causes the net to drop, with the fluttering victims safely secured in its meshes.

When the flight has passed, the women, girls, boys, and even the children, rush forth from their hiding-places; and, seizing the struggling birds, put a quick termination to their fruitless efforts, by biting each of them in the neck. Old, half-toothless crones — for this is especially their part of the performance — will be seen thus giving the final *coup* to the beautiful but unfortunate wanderers!

And still we have not explained what the boys are doing up yonder. Well, we shall now announce their *métier*. Each has taken up with him a number of little billets of wood, fashioned something like the letter Y, and about six inches in length. When this billet is flung into the air, and twirls about in its descent, it exhibits some resemblance — though not a very close one — to a flying pigeon-hawk. The resemblance, however, is near enough to "do" the pigeons; for when they are within about one hundred yards of the crows'-nest, the boy launches his billet into the air, and the birds, believing it to be a hawk, immediately dip several yards in their flight — as they may be seen to do when a real hawk makes his appearance. This descent usually brings them low enough to pass between the trees ; and of course the old women soon get their teeth upon them.

The pigeon-catching is not free to every one who may take a "fancy" to it. There are pigeon-catchers by trade ; who, with their families, follow it as a regular calling during the season, while it lasts ; and this, as already stated, is in the months of September and October. The *Palombière,* or pigeon-ridge, belongs to the communal authorities, who let it out in sections to the people that follow the calling of pigeon-netting ; and these, in their turn, dispose of the produce of their nets in the markets of Bagnères and other neighboring towns.

Every one knows how excellent for the table is the flesh of this beautiful bird : so much is it esteemed, that even at Bagnères, in the season of their greatest plenty, a pair will fetch a market price of from twelve to twenty sous.

CHAPTER XIX.

THE PYRENEES.

SPEAKING geologically, the Pyrenees extend along the whole north of Spain, from the Mediterranean to the province of Galicia on the Atlantic; and in this sense the chain may be regarded as between six and seven hundred miles in length. More properly, however, the term "Pyrenees" is limited to that portion of the range which lies directly between France and Spain; in other words, along the neck or isthmus of the Spanish peninsula. Thus limited, the range is less than half the above length, or about three hundred miles; while its average breadth is fifty.

Though less elevated than the Alps, the Pyrenees mountains are no molehills. Their highest peak, Maladetta, towers above 11,000 feet; and several others are of nearly equal height — while more than forty summits reach the elevation of 9,000!

The most elevated peaks are near the centre of the Pyrenees, the range gradually dipping downward as the extremities are approached. For this reason the most practicable passes are found near the eastern and western ends; though many also exist in the central part of the chain. In all, there are fifty passes or "ports," as they

are called, leading from the French to the Spanish side ;
but only five of these are practicable for wheeled vehi-
cles ; and a large number are only known (or at all
events only travelled) by the smugglers — *contrabandistas* — a class of gentry who swarm on both sides of the
Pyrenean frontier.

The superficial extent of these mountains is about
11,000 or 12,000 square miles.   Part of this is French,
and the remainder Spanish territory.   As a general rule,
the "divide," or main axis of the ridge forms the boundary line ; but in the eastern section the French terri
tory has been extended beyond the natural frontier.

The geological formation of the Pyrenees consists both
of primitive and secondary rocks, — the latter being
greater in mass, and composed of argillaceous schist,
grauwacke (schistose and common), and limestone.
Mines of lead, iron, and copper are found in this formation — the lead containing a proportion of silver.   The
primitive rocks are granite ; and run in zones or belts,
extended lengthwise in the direction of the chain ; and it
is in the rupture between these and the transition strata
that the chemical springs, for which the Pyrenees are so
famous, gush forth.   Of these remarkable fountains —
many of them almost at boiling heat — no less than 253
have been discovered in different parts of the range.   A
great number of them are celebrated for their medicinal
virtues, and are the favorite summer resorts of invalids,
as well as the votaries of pleasure, from all parts of the
world — but more especially from France and Spain.

The botany of the Pyrenees is full of interest.   It
may be regarded as an epitome of the whole European

flora : since scarcely a plant exists, from the Mediterranean to the Arctic Sea, that has not a representative species in some part of this mountain chain. In the valleys and lower slopes of the mountains the forest is chiefly composed of Lombardy poplars and sycamores ; a little higher, the Spanish chestnut, oaks, hazels, and alders, the mountain-ash and birch-trees abound ; and still farther up you enter the region of the pines — the *pinus sylvestris* growing in dense continuous forests, while the more graceful " stone-pine " is seen only in isolated groups or scattered trees. Everywhere a rich *flora* meets the eye ; flowers of the most lovely hues reflected in crystal rivulets — for the waters of the Pyrenees are pure beyond comparison, such a thing as a turbid stream being unknown throughout the whole range.

Above the pine forests the mountains exhibit a zone of naked declivities, stretching upward to the line of congelation — which in the Pyrenees is higher than upon the Alps. The former has been variously estimated : some fixing it at 8,300 feet, while others raise it as high as 9,000 ; but indeed, it would be more just to say that the snow-line depends greatly upon the locality of the particular mountain, and its southern or northern exposure.

In any case, it is more than 1,000 feet higher than on the Alps ; the superior elevation being accounted for, by the more southern latitude of the Franco-Spanish chain. Perhaps the proximity of the sea has more to do with this phenomenon than the trifling difference of latitude.

Upon the higher declivities and summits, snow-fields and glaciers abound, as in the Alps ; and even in some of the passes these phenomena are encountered. Most of the passes are higher than those of the Alps; but in consequence of the greater elevation of the snow-line, they remain open throughout the winter. At all seasons, however, they are by no means easy to traverse; and the cold winds that whistle through them are scarce to be endured. The Spaniards, who have a proverbial expression for almost every idea, have not neglected this one. In the ports (*puertos*) of the Pyrenees, say they, "the father waits not for his son, **nor the son** for his father."

If the passes across these mountains are higher than those of the Alps, the transverse valleys **are** the reverse; those of the Pyrenees being in general much lower. The consequence is, that from the bottom of these valleys the mountains themselves appear far loftier than any of the Alpine peaks, — the eye taking in at one **view a** greater angle of elevation.

The *fauna* of the Pyrenean chain, though less full and varied than its *flora,* is nevertheless of great interest. In the more densely wooded solitudes, and higher declivities of the mountains, a large bear is found, whose light fulvous-colored body and black paws pronounce him a different animal from the *ursus arctos.* If he be the same species, **as** naturalists assert, he **claims at** least to **be** a permanent variety, and deserves his distinctive appellation of *ursus pyrenaicus.*

Wolves abound; Spanish wolves, long famed for their fierceness; the common whitish-brown wolf (*canis lupus*), and a darker and still larger variety — in short, a black wolf, designated the "wolf of the Pyrenees," though it is equally a denizen of the other mountain *sierras* of Portugal **and** Spain.

The European lynx (*felis lynx*), and the wild-cat, both skulk through the Pyrenean forests; the former now only rarely seen. Along the naked cliffs leaps the " izzard," which is identical with the chamois of the Alps (*antelope rupicapra*); and in the same localities, but more rarely seen, the " bouquetin," or " tur " (*aigocerus pyrenaicus*) — a species of ibex, *not* identical with the *capra ibex* of Linnæus and the Alpine mountains.

Birds of many European species frequent the lower forests of the Pyrenees, or fill the sheltered valleys with their vocal music; while, soaring above the mountain summits, may be seen the great vulture-eagle, or " lammergeyer," watching with greedy eye the feeble lambkin, or the new-born kid of the ibex and izzard.

With such knowledge of their natural history, it was with feelings of no ordinary interest that our young hunters turned their faces towards that vast serried rampart that separates the land of the Gaul from the country of the Iberian.

It was by the Val d'Ossau, literally the " valley of the bear," that they made their approach to the mountains, — that valley celebrated as the residence and hunting-ground of Henri of Navarre; but now, in modern days, noted for its valuable thermal springs of *Eaux Bonnes* and *Eaux Chaudes*.

Up this mountain gorge went our heroes, their faces turned southward, and their eyes carried high up to the Pic du Midi d'Ossau — the mountain of the bears — an appropriate name for that beacon which was now directing their course.

CHAPTER XX.

AN ODD AVALANCHE.

IT is needless to say that the young Russians were delighted with the scenes that met their eyes in this fair southern land; and many of them **are** found faithfully described in their journal. They noted the picturesque dresses of the Pyrenean peasantry — so different from **the** eternal blue blouse which they had met in northern and central France. Here was worn the "barret," of scarlet or white, the rich brown jacket and red sash of the peculiar costumes of the Basque and Béarnais peasants — a fine race of men, and one, too, historically noble. **They** saw **carts** drawn by large-limbed cream-colored oxen; and passed flocks of sheep and milch goats, tended by shepherds in picturesque dresses, and guarded by numbers of large Pyrenean dogs, whose chief duty was to protect their charge from the wolves. They saw men standing knee-deep in the water, surrounded by **droves of** pigs — **the** latter voluntarily submitting themselves to **a** process of washing, which resulted in producing over their skins a roseate, pinky appearance. It could be seen, too, that these *pachyderms* not only submitted voluntarily to the operation, but with **a** keen sense of enjoyment, as evinced by their contented grunts, and by their long tails hanging "kinkless" while the large calabashes of water were poured over their backs. Perhaps to this careful management of the Pyrenean pigs are the beautiful "Bayonne hams" indebted for their celebrity.

Further on, our travellers passed a *plumire*, or "hen-bath." Here was a tank — another thermal spring — in which the water was something more than "tepid." In fact, it was almost on the boil; and yet in this tank a number of women were ducking their hens — not, as might be supposed, dead ones, in order to scald off their feathers, but live fowls, to rid them, as they said, of parasitical insects, and make them feel more comfortable! As the water was almost hot enough to *parboil* the poor birds, and as the women held them in it immersed to the necks, the *comfort* of the thing — so thought our travellers — was rather a doubtful question.

A little further on, another "custom" of the French Pyrenees came under the eyes of the party. Their ears were assailed by a singular medley of sounds, that rose from a little valley near the side of the road. On looking into the valley, they saw a crowd of forty or fifty women, all engaged in the same operation, which was that of flax-hackling. They learnt from this that in the Pyrenean countries the women are the hacklers of flax; and that, instead of each staying at her own home to perform the operation, a large number of them meet together in some shaded spot, bringing their unhackled flax along with them; and there, amidst jesting and laughing and singing, the rough staple is reduced to its shining and silky fineness.

Still another curious custom was observed ; but this was further on, and higher up the sides of the mountains. Their observation of it was attended with some degree of danger, and therefore came very close on being an " adventure." **For** this reason it found a place among the events recorded in their journal.

It should be remarked, that all three were mounted — Alexis and Ivan upon stout, active ponies, of that race for which the Pyrenees — especially the western section of them — are celebrated. Pouchskin's mount was not of the genus *equus*, **nor** yet an *asinus*, but a hybrid of both genera, — in short, a mule.

It was a French mule, and a very large one : **for it** required a good-sized quadruped of the kind to make an appropriate roadster for the ex-grenadier of the Imperial guard. It was not a very fat mule, however, but rawboned and gaunt as a Pyrenean wolf.

Of course these animals were all hired ones — obtained at Eaux Bonnes, and engaged for the trip across the Pyrenees to the Spanish side — as also to be used in any deviations that the hunters should think proper to make, while engaged in the pursuit of the bear.

From the nearest village on the Spanish side, the animals were to be sent back to their owner ; for it was not the intention of our travellers to return to the French territory.

Having crossed the mountains, and accomplishing the object for which they had visited them, their course would then be continued southward through Spain.

Along with them — also mounted on mule-back — was a fourth individual, whose services they had secured.

His *métier* was manifold — on this occasion combining in his single person at least three purposes. First, he was to serve them as guide; secondly, he was to bring back the hired horses; and, thirdly, he was to aid them in the "chasse" of the bear: for it so happened that this man-of-all-work was one of the most noted "izzard-hunters" of the Pyrenees. It is scarcely correct to say it *happened* so. Rather was it a thing of design than chance; for it was on account of his fame as a hunter, that he had been engaged for the triple duty he was now called upon to fulfil.

The four travellers, then, all mounted as we have described, were ascending a very steep declivity. They had left the last hamlet — and even the last house — behind them; and were now climbing one of the outlying spurs that project many miles from the main axis of the mountains. The road they were following scarcely deserved the name; being a pack-road, or mere bridle-path; and so steep was the ascent, that it was necessary to zigzag nearly a dozen times, before the summit of the ridge could be attained.

While entering upon this path, and near the base of the ridge, they had noticed the forms of men far above them, moving about the summit, as if engaged in some work. Their guide told them that these men were faggot-cutters, whose business was to procure firewood for the towns in the valley.

There was nothing in this bit of information to produce astonishment. What *did* astonish our travellers, however, was the mode in which these men transported their firewood down the mountain, of which, shortly after, they

were treated to an exhibition. As they were zigzagging
up the mountain-path, their ears were all at once saluted
by a noise that resembled a crashing of stones, mingled
with a crackling of sticks. The noise appeared to proceed from above ; and, on looking up, they beheld a
number of dark objects coming in full rush down the
declivity. These objects were of rounded form — in fact,
they were bundles of faggots — and so rapidly did they
roll over, and make way down the mountain, that had our
travellers chanced to be in their track, they might have
found some difficulty in getting **out of the** way.

Such was their reflection **at the** moment ; and they
were even thanking their stars that they had escaped the
danger, when all at once a fresh avalanche of faggots
was launched from above ; and these were evidently
bounding straight towards the party ! It was impossible
to tell which way to go — whether to rush forward or
draw back : for what with the inequality of the mountain-side, and the irregular rolling of the bundles, they
could not tell the exact direction they would take. All
therefore **drew** up, **and** waited the result in silent apprehension. Of course they had not long to wait — scarce a
second — for the huge bundles bounding on, each moment
with increased impetus, came down with the suddenness
of a thunder-clap ; and before the words " Jack Robin-
son " could have been pronounced, they went whizzing
past with the velocity of aërolites, and with such a force,
that had one of them struck either mule or pony it would
have hurled both the quadruped and its rider to the bottom of the mountain. It was only their good fortune
that saved them : for in such a place it would have been

impossible for the most adroit equestrian to have got out of the way. The path was not the two breadths of a horse; and to have wheeled round, or even drawn back upon it, would have been a risk of itself.

They rode on, again congratulating themselves on their escape; but fancy their consternation when they found themselves once more, and for the third time, exposed to the very same danger! Again came a set of bundles rolling and tearing down the slope, the billets rattling and crackling as they rolled; again they went swishing by; again, by the merest accident, did they miss the travellers, as they stood upon the path.

Now, it might be supposed that the faggots were being launched all along the ridge of the hill; and that, go which way they might, our party would still be exposed to the danger. Not so. The bundles were all rolled down at one particular place — where the slope was most favorable for this purpose — but it was the zigzag path, which every now and then obliqued across the line of the wood-avalanche, that had thus repeatedly placed them in peril.

As they had yet to "quarter" the declivity several times before they could reach the summit, they were more careful about approaching the line of descent; and whenever they drew near it, they put their ponies and mules to as good a speed as they could take out of them.

Though all four succeeded in reaching the summit in safety, it did not hinder Pouchskin from pouring out his vial of wrath on the heads of the offending wood-cutters; and if they could have only understood his Russian, they

would have heard themselves called by a good many
hard names, and threatened with a second pursuit of Mos-
cow. " Frog-eating Frenchmen ! " was the very mildest
title which the ex-guardsman bestowed upon them ; but
as his Russian was not translated, of course the phrase
fell harmless — else it would have undoubtedly been
retaliated by a taunt about " tallow."

The " izzard-hunter " swore at them to more purpose ;
for he, too, having undergone equal risk with the rest of
the party, had equally good reasons for being angry ; and
giving utterance to a long string of execrations with all
the volubility of a Béarnais, he further threatened them
with the terrors of the law.

As the wood-cutters, slightly stupefied by this unexpected attack, submitted with tolerable grace, and said
nothing in reply, the izzard-hunter at length cooled down,
and the party proceeded on their way ; Pouchskin, as he
rode off, shaking his clenched fist at the staring log-chop-
pers, and hissing out in angry aspirate another Russian
shibboleth, which neither could nor should be translated.

CHAPTER XXI.

A MEETING WITH MULETEERS.

A LITTLE beyond the scene of their encounter with the wood-cutters, the path entered among the gorges of the mountains, and the level plains of France were for a time lost to their view. The route they were following was a mere bridle-track, quite impracticable for carriages, but leading to one of the "ports" already mentioned, by which they could pass through to the Spanish side. Through this port a considerable traffic is carried on between the two countries — most of the carrying being done by Spanish muleteers, who cross the mountains conducting large trains of mules — all, except those upon which they themselves ride, laden with packs and bales of merchandise.

That such a traffic was carried over this route, our Russian travellers needed no other evidence than what came under their own eyes ; for shortly after, on rounding a point of rock, they saw before them a large drove of mules, gayly caparisoned with red cloth and stamped leather, and each carrying its pack. The gang had halted on a platform of no great breadth ; and the drivers — about a dozen men in all — were seen seated upon the rocks, a little way in advance of the animals. Each

6

wore a capacious cloak of brown cloth — a favorite color
among the Pyrenean Spaniards; and what with their
swarthy complexions, bearded lips, and wild attire, it
would have been pardonable enough to have mistaken
them for a band of brigands, or, at all events, a party
of *contrabandistas*.

They were neither one nor the other, however; but
honest Spanish muleteers, on their way to a French
market, with commodities produced on the southern
side of the mountains.

As our travellers came up, they were in the act of
discussing a luncheon, which consisted simply of black
bread, tough goat's-milk cheese, and thin Malaga wine
— the last carried in a skin bag, out of which each individual drank in his turn, simply holding up the bag and
pouring the wine by a small jet down his throat.

They were good-humored fellows, and invited our travellers to taste their wine; which invitation it would have
been ill-mannered to refuse. Ivan and Alexis emptied
some out into their silver cups — which they carried
slung conveniently to their belts; but Pouchskin, not
having his can so ready, essayed to drink the wine after
the fashion of the muleteers. But the goat-skin bag,
clumsily manipulated in the hands of the old guardsman, instead of sending the stream into his mouth, jetted
it all over his face and into his eyes, blinding and half-
choking him! As he stood in his stultified attitude, wine-
skin in hand, the precious fluid running down his nose,
and dripping from the tips of his grand mustachios, he
presented a picture that caused the muleteers to laugh
till the tears ran down their cheeks; shouting out their

bravos and other exclamations, as if they were applauding
some exquisite piece of performance in a theatre.

Pouchskin took it all in good part, and the muleteers
pressed him to try again ; but, not caring to expose himself to a fresh burst of ridicule, the old grenadier borrowed the cup of one of his young masters ; and by the
help of this managed matters a little more to his mind.
As the wine tasted good to the old soldier's palate, and as
the hospitable muleteers invited him to drink as much as
he pleased, it was not until the goat-skin bag exhibited
symptoms of collapse, that he returned it to its owners.

Perhaps had Pouchskin not indulged so freely in the
seducing Malaga tipple, he might have avoided a very
perilous adventure which befell him almost on the instant, and which we shall now relate.

Our travellers, after exchanging some further civilities
with the muleteers, had once more mounted, and were
about proceeding on their way. Pouchskin, riding his
great French jennet, had started in the advance. Just in
front of him, however, the pack mules were standing in a
cluster — not only blocking up the path, but barring the
way on both sides — so that to get beyond them it would
be necessary to pass through their midst. The animals
all seemed tranquil enough, — some picking at the bushes
that were within their reach, but most of them standing
perfectly still, occasionally shaking their long ears, or
changing one leg to throw the weight upon another.
Pouchskin saw that it was necessary to pass among them ;
and, probably, had he squeezed quietly through, they
might have remained still, and taken no notice of him.
But elated with the wine he had drunk, the ex-grenadier,

instead of following this moderate course, drove his spurs
into his great French hybrid, and with a loud charging
yell — such as might have issued from the throat of a
Cossack — he dashed right into the midst of the drove.

Whether it was because the animal he **bestrode** was
French, or whether something in Pouchskin's **voice** had
sounded ill in their ears, it is not possible to say, but all
at once the whole Spanish *mulada* was perceived to be in
motion, — each individual mule rushing toward Pouchskin with pricked ears, **open** mouth, and tail elevated in
the air ! It was too late for **him** to hear the cry of the
izzard-hunter, "*prenez-garde !*" or the synonyme, "*guarda
te !*" of the muleteers. He may have heard both these
cautionary exclamations, but they reached him too late
to be of any service to him : for before he could have
counted six, at least twice that number of mules **had**
closed round him, and with a simultaneous scream commenced snapping and biting at both him and his French
roadster **with all the** fury of famished wolves ! In vain
did the stalwart jennet defend **itself** with **its** shod hoofs,
in vain did its rider lay round him with his whip : for not
only did the Spanish mules assail him with their teeth,
but, turning tail as well, they sent their heels whistling
around his head, and now and then thumping against
his legs, until **his** leather boots and breeches cracked
under their kicks !

Of course, the muleteers, on perceiving Pouchskin's
dilemma, had rushed instantaneously to the rescue ; and,
with loud cries, and cracking of their whips, — as muleteers alone can crack them, — were endeavoring to beat
off the assailants. But, with all their exertions, backed

by their authority over the animals, Pouchskin might
have fared badly enough, had not an opportunity offered
for extricating himself. His animal, fleeing from the
persecution of its Spanish enemies, had rushed in among
some boulders of rock. Thither it was hotly pursued;
and Pouchskin would again have been overtaken, had he
not made a very skilful and extensive leap out of the
saddle, and landed himself on a ledge of rock. From
this he was able to clamber still higher, until he had
reached a point that entirely cleared him of the danger.

The French jennet, however, had still to sustain the
attack of the infuriated mules; but, now that it was relieved from the encumbrance of its heavy rider, it gained
fresh confidence in its long legs; and making a dash
through the midst of the *mulada*, it struck off up the
mountain-path, and galloped clear out of sight. The
mules, encumbered with their packs, did not show any
inclination to follow, and the drama was thus brought to
a termination.

The woebegone look of the old guardsman, as he stood
perched upon the high pinnacle of rock, was again too
much for the muleteers; and one and all of them gave
utterance to fresh peals of laughter. His young masters
were too much concerned about their faithful Pouckskin
to give way to mirth; but on ascertaining that he had
only received a few insignificant bruises, — thanks to the
Spanish mules not being shod, — they, too, were very
much disposed to have a laugh at his expense. Alexis
was of opinion that their follower had made rather free
with the wine-skin; and therefore regarded the chastise-
ment rather in the light of a just retribution.

It cost the izzard-hunter a chase before Pouchskin's
runaway could be recovered; but the capture of the jennet was at length effected; and, all things being set to-
rights, a parting salute was once more exchanged with
the muleteers, and the travellers proceeded on their
way.

CHAPTER XXII.

THE PYRENEAN BEARS.

It was well they had the izzard-hunter for a guide, for without him they might have searched a long time without finding a bear. These animals, although plenteous enough in the Pyrenees some half-century ago, are now only to be met with in the most remote and solitary places. Such forest-tracts as lie well into the interior gorges of the mountains, and where the lumberer's axe never sounds in his ears, are the winter haunts of the Pyrenean bear; while in summer he roams to a higher elevation — along the lower edge of the snow-fields and glaciers, where he finds the roots and bulbs of many Alpine plants, and even lichens, congenial to his taste. He sometimes steals into the lower valleys, where these are but sparsely cultivated; and gathers a meal of young maize, or potatoes, where such are grown. Of truffles, he is as fond as a Parisian sybarite, — scenting them with a keenness far excelling that of the regular truffle-dog, and " rooting " them out from under the shade of the great oak-trees, where these rare delicacies are inexplicably produced.

Like his near congener, the brown bear, he is frugivorous; and, like most other members of their common

family, he possesses a sweet tooth, and will rob bees of their
honey whenever he can find a hive. He is carnivorous
at times, and not unfrequently makes havoc among the
flocks that in summer are fed far up on the declivities
of the mountains; but it has been observed by the shepherds that only odd individuals are given to this san-
guinary practice, and, as a general rule, the bear will
not molest their sheep. On this account, a belief exists
among the mountaineers that there are two kinds of
bears in the Pyrenees; one, an eater of fruits, roots, and
larvæ, — the other, of more carnivorous habits, that eats
flesh, and preys upon such animals as he can catch. The
latter they allege to be larger, of more fierce disposition,
and when assailed, caring not to avoid an encounter with
man. The facts may be true, but the deduction erroneous. The izzard-hunter's opinion was that the Pyrenean
bears were all of one species; and that, if there were
two kinds, one was a younger and more unsophisticated
sort, the other a bear whom greater age has rendered
more savage in disposition. The same remark will
apply to the Pyrenean bear that is true of the *ursus
arctos*, — viz. having once eaten flesh, he acquires a taste
for it; and to gratify this, of course the fiercest passions
of his nature are called into play. Hunger may have
driven him to his first meal of flesh meat; and afterwards he seeks it from choice.

The izzard-hunter's father remembered when bears
were common enough in the lower valleys; and then not
only did the flocks of sheep and goats suffer severely,
but the larger kinds of cattle were often dragged down
by the ravenous brutes — even men lost their lives in

encounters with them! In modern times, such occurrences were rare, as the bears kept high up the mountains, where cattle were never taken, and where men went very seldom. The hunter stated that the bears were much sought after by hunters like himself, as their skins were greatly **prized, and** fetched a good price; that the young bears were also very valuable, and to capture a den of **cubs was esteemed** a bit of rare good luck: since these were brought up to be used in the sports of bear-baiting **and** bear-dancing, spectacles greatly relished in the frontier towns of France.

He knew of no particular mode for taking bears. Their chase was too precarious to make it worth while; and they were only encountered accidentally by the izzard-hunters, when in pursuit of their own regular game. Then they were killed by being shot, if old ones; and if young, they captured them by the aid of their dogs.

" So scarce are they," added the hunter, "that I have killed only three this whole season; but I know where there's a fourth — a fine fellow, too ; and if you feel inclined —— "

The young Russians understood the hint. Money is all-powerful everywhere ; and a gold coin will conduct to the den of a Pyrenean bear, where the keenest-scented hound or the sharpest-sighted hunter would fail to find it. In an instant almost, the bargain was made. Ten dollars for the haunt of the bear !

The *Pic du Midi* **d'Ossau** was now in sight; and, leaving the beaten path that passed near its base, our hunters turned off **up a** lateral ravine. The sides and bottom of this ravine were covered with a stunted growth

of pine-trees; but as they advanced further into it, the
trees assumed greater dimensions — until at length they
were riding through a tall and stately forest. It was, to
all appearance, as wild and primitive as if it had been on
the banks of the Amazon or amid the Cordilleras of the
Andes. Neither track nor trail was seen — only the
paths made by wild beasts, or such small rodent animals
as had their home there.

The izzard-hunter said that he had killed lynxes in
this forest; and at night he would not care to be alone
in it, as it was a favorite haunt of the black wolves.
With such company, however, he had no fear; as they
could kindle fires and keep the wolves at bay.

The neighborhood in which he expected to find the
bear was more than two miles from the place where they
had entered the forest. He knew the exact spot where
the animal was at that moment lying — that is, he knew
its cave. He had seen it only a few days before, going
into this cave; but as he had no dogs with him, and no
means of getting the bear out, he had only marked the
place, intending to return, with a comrade to help him.
Some business had kept him at Eaux Bonnes, till the arrival of the strangers; and learning their intentions, he
had reserved the prize for them. He had now brought
his dogs — two great creatures they were, evidently of
lupine descent — and with these Bruin might be baited
till he should come forth from his cave. But that plan
was only to be tried as a last resource. The better
way would be to wait till the bear started out on his midnight ramble, — a thing he would be sure to do, — then
close up the mouth of the cave, and lie in ambush for

his return. He would "not come home till morning," said the izzard-hunter; and they would have light to take aim, and fire at him from their different stations.

It seemed a feasible plan, and as our adventurers now placed themselves in the hands of the native hunter, it was decided they should halt where they were, kindle a fire, and make themselves as comfortable as they could, until the hour when Bruin might be expected to go out upon his midnight prowl.

A roaring fire was kindled; and Pouchskin's capacious haversack being turned inside out, all four were soon enjoying their dinner-supper with that zest well known to those who have ridden twenty miles up a steep mountain-road.

CHAPTER XXIII.

THE IZZARD-HUNTER.

THEY passed the time pleasantly enough, listening to
the stories of the izzard-hunter, who related to them
much of the lore current among the peasantry of the
mountains — tales of the chase, and of the contraband
trade carried on between Spain and France, besides
many anecdotes about the Peninsular war, when the
French and English armies were campaigning in the
Pyrenees. In this conversation Pouchskin took part:
for nothing was of greater interest to the old soldier than
souvenirs of those grand times, when Pouchskin entered
Paris.

The conversation of the izzard-hunter related chiefly
to his own calling, and upon this theme he was enthusiastic. He told them of all the curious habits of the
izzard; and among others that of its using its hooked
horns to let itself down from the cliffs — a fancy which
is equally in vogue among the chamois-hunters of the
Alps, but which Alexis did not believe, although he did
not say so — not wishing to throw a doubt on the veracity of their guide. The latter, however, when closely
questioned upon the point, admitted that he had never
himself been an eyewitness of this little bit of goat gym-

nastics; he had only heard of it from other hunters, who said they had seen it; and similar, no doubt, would be the answer of every one who spoke the truth about this alleged habit of the chamois. The fact is, that this active creature needs no help from its horns. Its hoofs are sufficient to carry it along the very narrowest ledges; and the immense leaps it can take either upward or downward, can be compared to nothing but the flight of some creature furnished with wings. Its hoof, too, is sure, as its eye is unerring. The chamois never slips upon the smoothest rocks — any more than would a squirrel upon the branch of a tree.

Our travellers questioned the izzard-hunter about the profits of his calling. They were surprised to find that the emolument was so trifling. For the carcass of an izzard he received only ten francs; and for the skins two or three more! The flesh or venison was chiefly purchased by the landlords of the hotels — of which there are hundreds at the different watering-places on the French side of the Pyrenees. The visitors were fond of izzard, and called for it at the table. Perhaps they did not relish it so much as they pretended to do; but coming from great cities, and places where they never saw a chamois, they wished to be able to say they had eaten of its flesh. In this conjecture the izzard-hunter was, perhaps, not far out. A considerable quantity of game of other kinds is masticated from a like motive.

It was suggested by Ivan, that, with such a demand for the flesh, the izzard should fetch a better price. Ten francs was nothing.

"Ah!" replied the hunter with a sigh, "that is easily

explained, monsieur! The hotel-keepers are too cunning, both for us and their guests. If we were to charge more, they would not take it off our hands."

"But they would be under the necessity of having it, since their guests call for it."

"So they do; and if there were no *goats*, our izzard-venison would sell at a higher price."

"How?" demanded Ivan, puzzled to make out the connection between goats and izzard-venison.

"Goats and izzards are too much alike, monsieur — that is, after being skinned and cut up. The hotel-keeper knows this, and often makes 'Nanny' do duty for izzard. Many an hotel traveller at Eaux Bonnes may be heard praising our izzard's flesh, when it is only a quarter of young kid he's been dining upon. Ha! ha! ha!"

And the hunter laughed at the cheat — though he well knew that its practice seriously affected the income of his own calling.

But, indeed, if the truth had been told, the man followed the chase far less from a belief in its being a remunerative profession, than from an innate love for the hunter's life. So enthusiastic was he upon the theme, that it was easy to see he would not have exchanged his calling for any other — even had the change promised him a fortune! It is so with professional hunters in all parts of the world, who submit to hardships, and often the greatest privations, for that still sweeter privilege of roaming the woods and wilds at will, and being free from the cares and trammels that too often attach themselves to social life.

Conversing on such topics, the party sat around the

bivouac fire until after sunset, when their guide admonished them that they would do well to take a few hours of sleep. There was no necessity for going after the bear until a very late hour — that is, until near morning — for then the beast would be most likely to be abroad. If they went too soon, and found him still in his cave, it was not so certain that even the dogs could prevail on him to turn out. It might be a large cavern. He might give battle to the dogs inside; and big as they were, they would be worsted in an encounter of that sort: since a single blow from the paw of a bear is sufficient to silence the noisiest individual of the canine kind. The dogs — as the hunter again repeated — should only be used as a last resource. The other plan promised better; as the bear, once shut out of his cave, would be compelled to take to the woods. The dogs could then follow him up by the fresh scent; and unless he should succeed in finding some other cavern in which to ensconce himself, they might count upon coming up with him. It was not uncommon for the Pyrenean bear, when pursued by dogs and men, to take to a tree; and this would be all that their hearts could desire: since in a tree the bear would be easily reached by the bullets of their guns. Besides, they might have a chance, when he returned to his closed cave, to shoot him down at once; and that would end the matter without further trouble.

It was not necessary to go to the cave until near morning — just early enough to give them time to close up the entrance, and set themselves in ambush before day broke. On this account the guide recommended them to take some sleep. He would answer for it that they should be waked up in time.

This advice was cheerfully accepted and followed. Even Pouchskin required repose, after the rough handling he had received at the *mouths* of the mules; and he was now quite as ready as his young masters to wrap himself up in his ample grenadier great-coat, and surrender himself into the arms of the Pyrenean Morpheus.

CHAPTER XXIV.

THE AMBUSCADE.

TRUE to his promise, the izzard-hunter awoke them about an hour before dawn; and having saddled and bridled their animals, they mounted and rode off. Among the great tree-trunks it was very dark; but the hunter knew the ground; and, after groping along for half a mile farther, and somewhat slowly, they arrived at the base of a cliff. Keeping along this for some distance farther, they came at length to the place of their destination — the mouth of the cave. Even through the gloom, they could see a darker spot upon the face of the rock, which indicated the entrance. It was of no great size — about large enough to admit the body of a man in a stooping attitude — but the hunter was under the impression that it widened inward, and led to a grand cavern. He drew his inference, not from having ever explored this particular cave, but from knowing that there were many others of a similar kind in that part of the mountains, where the limestone formation was favorable to such cavities. Had it been only a hole just big enough for the den of a bear, he would have acted very differently — then there would have been a hope of drawing Bruin out with the dogs; but if the place was an actual cavern, where the beast might range freely about, the hunter knew there would be no chance of getting him out. Their presence outside once suspected, the bear might remain for days within his secure fortress; and a siege would have to be laid, which would be a tedious affair, and might prove fruitless in the end.

For this reason, **great** caution had been observed as they drew near the cave. They feared that they might come upon the bear, by chance wandering about in the woods,— that he might hear them, and, taking the alarm, scamper back to his cavern.

Acting under this apprehension, they had left their animals a good way off — having tied them to the trees — and had approached the cave on foot, without making the slightest noise, and talking to each other only in whispers.

The izzard-hunter now proceeded to put his designs into execution. While the others had been sleeping, he **had** prepared a large torch, out of dry splinters of the stone-pine; and now quietly igniting this, set it in the ground near the base of the cliff. The moment the bright flame illuminated the entrance to the cave, all stood with their guns in hand ready to fire. They were not sure that Bruin had gone out at all. He might still be abed. **If so,** the light of the torch might wake him up and tempt him forth; therefore it was best to be prepared for such a contingency.

The izzard-hunter now slipped his dogs, which up to this time he had held securely in the leash. As soon as they were free, the well-trained animals, knowing what was expected of them, rushed right into the cave.

For some seconds the dogs kept up a quick continuous yelping, and their excited manner told that they at least scented a bear: but the question to be determined was, whether the brute was still in his den.

The hunter had surmised correctly. The aperture conducted to a real cavern, and a very large one — as could be told by the distance at which the yelping of the dogs was heard. Out of such a place it would have been hopeless to have thought of starting a bear — unless it should please Bruin to make a voluntary exit. It was, therefore, with no little anxiety that the hunters listened to the "tongue" of the dogs, as it echoed within the cavernous hollow.

They all knew that if the bear should prove to be inside, the dogs would soon announce the fact by their barking, and other fierce sounds characteristic of canine strife.

They were not kept long in suspense; for, after an interval of less than a minute, both dogs came running out, with that air of disappointment that told of their having made an idle exploration.

Their excited movements, however, proved that the scent of the bear was fresh, — that he had only recently forsaken his den, — for the dogs had been heard scratching among the sticks and grass that composed it; but this only showed clearly that his habitation was untenanted, and Bruin was "not at home."

This was just what the izzard-hunter desired; and all of them laying aside their guns, proceeded to close up the entrance. This was an easy task. Loose boulders lay around, and with these a battery was soon built across the mouth of the cavern, through which no animal could possibly have made an entrance.

The hunters now breathed freely. They felt certain they had cut **off** the retreat of the bear; and unless he should suspect something wrong, and fail to return to his cave, they would be pretty sure of having a shot at him.

Nothing remained but to place themselves in ambush, and wait for his coming. How to conceal themselves became the next consideration. It was a question, too, of some importance. They knew **not** which way the **bear** might come. He might see them while approaching, and trot off again before **a shot** could be fired. To prevent **this some extraordinary** measure must **be** adopted.

A plan soon presented itself **to** the practised hunter of the Pyrenees. Directly **in** front of the cliff grew several large trees. They **were** of the *pinus sylvestris*, and thickly covered with bunches of long needle-shaped leaves. **If** they should climb into these trees, the leaves and branches would sufficiently conceal them, and the bear would **hardly suspect** their presence in such a situation.

The suggestion of their guide was at once acted upon. Ivan and Pouchskin got into one tree, while the izzard-hunter and Alexis chose another; and all having secured places where they could command a view of the walled-up entrance without being themselves seen, they waited for daylight and **the** coming back of the bear.

CHAPTER XXV.

A BEAR IN A BIRD'S-NEST.

For the light they had not long to wait. The day broke almost as soon as they had got well settled in their places; but the bear was likely to delay them a little longer — though how long it was impossible to guess, since his return to his sleeping-quarters might depend on many contingencies. Formerly the Pyrenean bears — so the izzard-hunter said — were often met with ranging about in the daytime; but that was when they were more numerous, and less hunted. Now that they were scarce, and their skins so highly prized — which, of course, led to their becoming scarcer every day, and more shy too — they rarely ever left their hiding-place except during the night, and in this way they contrived to escape the vigilance of the hunters. As to the one they were waiting for, the hunter said he might return earlier or later, according to whether he had been much chased of late.

The exact time of his return, however, was soon after ascertained, by the bear himself making his appearance right under their noses.

All at once, and in the most unexpected manner, the great quadruped came shuffling up to the mouth of the cave. He was evidently moving under some excitement, as if pursued, or alarmed by something he had seen in the woods. It was perhaps the sight of the horses, or else the scent of the hunters themselves — on whose track he appeared to have come. Whatever it was, the party in the trees did not take time to consider, or rather the bear did not give them time; for, the moment he reached the entrance to his cave, and saw that it was blocked up, he gave utterance to a terrific scream expressing disappointment, and turning in his tracks, bounded off, as rapidly as he had come up!

The volley of four shots, fired from the trees, caused some of his fur to fly off; and he was seen to stagger, as if about to fall. The hunters raised a shout of triumph, thinking they had been successful; but their satisfaction was short-lived : for, before the echoes of their voices died along the cliff, the bear seemed once more to recover his equilibrium, and ran steadily on.

Once or twice he was seen to stop, and face round to the trees — as if threatening to charge towards them; but again resigning the intention, he increased his speed, went off at a lumbering gallop, and was soon lost to their sight.

The disappointed hunters rapidly descended from their perch ; and letting loose the dogs, started off on the trail. Somewhat to their surprise, as well as gratification, it led near the place where they had left their animals; and as they came up to these, they had proofs of the bear having passed that way, by seeing all four, both ponies and mules, dancing about, as if suddenly smitten with madness. The ponies were " whighering," and the mules squealing, so that their owners had heard them long before coming in sight of them. Fortunately the animals had been securely fastened — else there was no knowing whither they would have galloped, **so** panic-stricken did they appear.

Our hunters believed it a fortunate circumstance that the bear had gone that way ; for the guide assured them that there was no telling where he would now stop ; and as the chase might carry them for miles through the mountains, they would have been compelled to take to their saddles before starting upon it. The direction the bear had taken, therefore, was just the one most convenient for his pursuers.

Staying no longer than to untie their animals, they once more mounted, and kept after the dogs, whose yelping they could hear already some distance in the advance.

As the izzard-hunter said, the Pyrenean bear, like his Norwegian cousin, when started from his lair, often scours the country to a great distance before making halt — not unfrequently deserting the ravine or mountain-side, where he habitually dwells, and making for some other place, where he anticipates finding greater security.

In this way he often puts his pursuers at fault — by passing over rocky shingle, along ledges of cliffs, or up precipitous slopes, where neither men nor dogs can safely follow him. This was just what they now had to fear ; for the guide well knew that the forest they were in was surrounded on almost every side by rocky cliffs ; and if the bear should get up these, and make to the bald mountains above, they would stand a good chance of losing him altogether.

But one hope the hunter had. He had perceived —
as indeed they all had — that several of their shots had
hit the bear — and that he must be severely wounded
to have staggered as he had done. **For** this reason he
might seek **a hiding-place in** the **forest, or** perchance take
to a tree. Cheered by **this** hope, the pursuers pushed
onward.

The conjecture proved to **be a** just one ; for before
they had gone half a mile farther, **a** continuous barking
sounded on their ears, which they **knew to** be that of the
dogs. They knew, moreover, by this sign, **that the** bear
had **done one of** three things — either **taken** to a tree,
retreated **into a cave, or come to a** stand in the open
ground, and was keeping the dogs at bay. Of the three
conjectures, they desired that the first should prove **the**
correct one ; and from the manner in which the dogs
were giving tongue, they **had** reason to hope that **it**
would.

In effect **so it** did ; for, on getting a little closer, the
two dogs were **seen** bounding about the roots of an enormous tree, at intervals springing up against its trunk, and
barking at some object that had taken refuge **in the**
branches above.

Of course, this object could **only** be **the** bear ; **and**
under this belief, **the** pursuers approached the tree —
each holding **his** gun cocked and ready **to fire.**

When **they had** got quite up to the tree, and **stood**
under it, no **bear was to** be seen ! A large black mass
was visible among the topmost branches ; but this was
not the body of a bear : it was something altogether different. The tree was one of gigantic size — the very

largest they had seen in the whole forest; it was a pine,
of the species *sylvestris*, with huge spreading limbs, and
branches thickly covered with fascicles of long leaves.
In many places the foliage was dark and dense enough to
have afforded concealment to an animal of considerable
size; but not one so bulky as a bear; and had there
been nothing else but the leaves and branches to conceal
him, a bear could not have found shelter in that tree
without being visible from below. And yet a bear was
actually in it — the very same bear they were in pursuit
of — though not a bit of his body — not even the tip of
his snout, was visible to the eyes of the hunters!

He was certainly there: for the dogs, who were not
trusting to their eyes, but to that in which they placed
far more confidence — their scent, — by their movements
and behavior, showed their positive belief that Bruin
was in the tree.

Perhaps you will fancy that the pine was a hollow one,
and that the bear had crept inside. Nothing of the kind:
the tree was perfectly sound — not even a knot-hole was
visible either in its trunk or limbs. It was not in a
cavity that Bruin had been able to conceal himself.

There was no mystery whatever about their not seeing
him: for as soon as the hunters got fairly under the tree,
and looked up, they perceived, amidst its topmost branches, the dark object already mentioned; and as the bear
could be seen nowhere else in the tree, this object accounted for his being invisible.

You will be wondering what it was; and so wondered
our young hunters when they first raised their eyes to it.
It looked more like a stack of faggots than aught else;

7

and, indeed, very good faggots would it have made : since
it consisted of a large mass of dry sticks and branches,
resting in an elevated fork of the tree, and matted together into a solid mass. There were enough to have
made a load for an ordinary cart, and so densely packed
together, that only around the edges could the sky be
seen through them ; towards the centre, and for a diameter as large as a millstone, the mass appeared quite solid
and black, not a ray of light passing through the inter-
woven sticks.

"The nest of a lammergeyer!" exclaimed the izzard-
hunter, the moment his eye glanced up to it. "Just so!
— my dogs are right : the bear has taken shelter in the
nest of the birds !"

CHAPTER XXVI.

THE LAMMERGEYERS.

THIS was evident to all. Bruin had climbed the tree, and was now snugly ensconced in the great nest of the vulture-eagles, though not a hair of his shaggy hide could be visible from below.

The hunters had no doubt about his being there. The *chasseur* was too confident in the instinct of his well-trained dogs to doubt them for a moment, and his companions had no reason to question a fact so very probable. Had there been any doubt, it would soon have been set aside, by an incident that occurred the moment after their arrival under the tree. As they stood looking upward, two great birds were seen upon the wing, rapidly swooping downward from on high. They were *lammer-qeyers,* and evidently the owners of the invaded nest. That the intruder was not welcome there, became apparent in the next moment ; for both the birds were seen shooting in quick curves around the top branches of the tree, flapping their wings over the nest, and screaming with all the concentrated rage of creatures in the act of being plundered. Whether Bruin, in addition to his unwelcome presence, had also committed burglary, and robbed the eagles of their eggs or young, could not be told. If he had done so, he could not have received greater objurgation from the infuriated birds, that continued their noisy demonstrations, until a shot fired from below admonished them of **the** presence of that biped enemy far more dreaded than the bear. Then did they only widen **the** circle of their flight, still continuing to swoop down **over** the nest at intervals, and uttering their mingled cries of rage and lamentation.

The shot was from the gun of the izzard-hunter; but it was not till after he had been some time upon the ground that he had fired it. All four had previously dismounted and fastened their animals **to the** surrounding trees. They knew that the bear was in the nest; but although his retreat was now cut off, it was still not so certain that they should succeed in making a **capture.** Had the bear taken refuge in a fork, or even among thick branches, where their bullets might have reached him, it would have been a very different thing. They might then have brought him down at their pleasure, for if killed, or severely wounded, he must have fallen to the ground; but now — ah, now! what was to be done? The broad platform of the nest not only gave him a surface on which he could recline at his ease, but its thick mass formed a rampart through which not even a bullet would be likely to penetrate to his body!

How were they to reach him with their bullets? That was the next question that came under consideration. The odd shot had been fired as an experiment. It was fired in the hope that it might startle the bear, and cause him to shift his quarters — if only a little — so that some part of his body might be exposed; and while the izzard-hunter was discharging his piece, the others had stood watching for a chance. None was given to them, however. The bullet was heard striking the sticks, and caused the dust to puff out, but it produced no further effect,—not a move was made by the occupant of that elevated eyrie.

Two or three more shots were fired with like effect; and the fusil of Pouchskin was next called into requisition, and brought to bear upon the nest. The large bullet crashed up among the dry sticks, scattering the fragments on all sides, and raising a cloud of dust that enveloped the whole top of the tree. But not a sign came from Bruin, to tell that it had disturbed him; not even a growl, to reward Pouchskin for the expenditure of his powder and lead. It was evident that this mode of proceeding could be of no service; and the firing was at once discontinued—in order that they might take into consideration some other plan of attack.

At first there appeared to be no way by which the bear might be ousted from his secure quarters. They might fire away until they should empty both their powder-horns and pouches, and all to no purpose. They might just as well fire their shots into the air. So far as their bullets were concerned, the bear might bid them defiance—a cannon-shot alone could have gone through his strong rampart of sticks.

What could they do to get at him? To climb up and assail him where he lay was not to be thought of—even could they have climbed into the nest. On the firm ground, none of them would have liked to risk an encounter with the enemy, much less upon such insecure footing as a nest of rotten sticks. But they could not have got into the nest, however bent upon such a thing. Its wide rim extended far beyond the supporting branches ; and only a monkey, or the bear himself, could have clambered over its edge. To a human being, ascent to the nest would have been not only difficult, but impossible ; and no doubt the instinct of the eagles guided them to this while they were constructing it. Not for a moment, then, did our hunters think of climbing up to their eyrie.

What, then, were **they to do?** The only thing they could think of was to cut down the tree. It would be a great undertaking : for the trunk was several feet in diameter ; and as they had only one axe, and that not a very sharp one, it would be a work of time. They might be days in felling that gigantic pine ; and even when down, the bear might still escape from them — for it did not follow that the fall of the tree would result in the consummation of his capture. It might swing over gradually and easily, or, striking against others, let the bear down without doing him the slightest damage ; and in the confusion consequent on its fall, he would have a good chance of getting off.

These considerations caused them to hesitate about cutting down the tree, and reflect whether there might not be some easier and more effective method for securing the skin of the bear.

CHAPTER XXVII.

FIRING THE EYRIE.

After beating their brains, for some time to no purpose, an exclamation from the izzard-hunter at length announced that some happy idea had occurred to him. All eyes were at once turned towards him; while the voice of Ivan was quickly heard, interrogating him as to the object of his exclamation.

"I've got a plan, young monsieur!" replied the hunter, "by which I'll either force the bear to come down, or roast him up yonder where he lies. *Parbleu!* I've got an excellent idea!"

"What is it? what is it?" eagerly inquired Ivan; though from what the izzard-hunter had said, he already half comprehended the design.

"Patience, young monsieur! in a minute you shall see!"

All three now gathered around the *chasseur*, and watched his movements.

They saw him pour a quantity of gunpowder into the palm of his hand; and then tear a strip of cotton rag from a large piece which he had drawn out of his pouch. This he saturated with saliva and then coated it over with the powder. He next proceeded to rub both rag

and powder together — until, after a considerable friction
between the palms of his hands, the cotton became once
more dry, and was now thoroughly saturated with the
powder, and quite blackened with it.

The next proceeding on the part of the chasseur was
to procure a small quantity of dead moss, which was
easily obtained from the trunks of the surrounding trees;
and this, mixed with a handful or two of dry grass he
rolled up into a sort of irregular clew.

The man now felt in his pouch; and, after a little
fumbling there, brought forth a small box that was seen
to contain lucifer-matches. Seemingly satisfied with their
inspection, he returned the box to its place, and then
made known the object for which all these preliminary
manœuvres had been practised. Our young hunters had
already more than half divined it, and it only confirmed
their anticipations when the hunter declared his intention
to climb the tree and set fire to the nest.

It is needless to say that one and all of them approved
of the scheme, while they admired its originality and cunning. Its boldness, too, did not escape their admiration,
for it was clearly a feat of daring and danger. The bottom of the nest might be reached easily enough; for
though a tall tree, it was by no means a difficult one to
climb. There were branches all along its trunk from
bottom to top; and to a Pyrenean hunter, who, when a
boy, as he told them, had played pigeon vidette in one of
the " crows'-nests " they had seen, the climbing of such a
tree was nothing. It was not in this that the danger lay,
but in something very different. It was in the contingency, that, while up in the branches, and before he

could effect his purpose, the bear might take a fancy to
come down. Should he do so, then, indeed, would the
life of the venturesome hunter be in deadly peril.

He made light of the matter, however, and, warning
the others to get their guns ready and stand upon their
guard, he sprang forward to the trunk, and commenced
" swarming " upward.

Almost as rapidly as a bear itself could have ascended,
the izzard-hunter glided up the tree, swinging himself
from branch to branch, and resting his naked feet — for
he had thrown off his shoes — on knots and other inequalities, where no branch offered. In this way he at
length got so close to the nest, that he could without difficulty thrust his hand into the bottom of it.

He was now seen drawing forth a number of the dry
sticks, and forming a cavity near the lower part of the
huge mass. He operated with great silence and circum-
spection — taking all the care he could not to make his
presence known to the bear, nor in any way disturb
whatever dreams or reflections Bruin might then be indulging in.

In a short time he had hollowed out a little chamber
among the sticks — just large enough for his purpose, —
and, taking the ball of dry grass out of his pouch, he
loosened it a little, and then placed it within the cavity.

It was but the work of another minute to light a luci-
fer-match, and set fire to the long strips of tinder rag
that hung downwards from the grass.

This done, the izzard-hunter swung himself to the next
branch below; and, even faster than he had gone up, he
came scrambling down the trunk.

7 *

Just as he reached the ground, the grass was seen
catching; and amidst the blue smoke that was oozing
thickly out of the little chamber, and slowly curling up
around the edges of the nest, a red blaze could be distinguished — accompanied with that crackling noise that
announces the kindling of a fire.

The four hunters stood ready, watching the progress
of the little flame — at the same time directing their
glances around the rim of the nest.

They had not long to wait for the *dénouement*. The
smoke had already caught the attention of the bear; and
the snapping of the dry faggots, as they came in contact
with the blazing grass, had awakened him to a sense of
his dangerous situation.

Long before the blaze had mounted near him, he has
seen craning his neck over the edge of the nest; first on
one side, then on another, and evidently not liking what
he saw. Once or twice he came very near having a
bullet sent at his head; but his restlessness hindered
them from getting a good aim, and for the time he was
left alone.

Not for long, however: for he did not much longer
remain upon his elevated perch. Whether it was the
smoke that he was unable longer to endure, or whether
he knew that the conflagration was at hand, does not
clearly appear; but from his movements it was evident
the nest was getting too hot to hold him.

And no doubt it was too hot at that crisis. Had he
remained in it but two minutes longer, an event would
have occurred that would have ruined everything. The
bear would either have been roasted to a cinder; or, at

all events, his skin would have been singed, and, of course, completely spoilt for the purpose for which it was required!

Up to this moment that thought had never occurred to the young hunters; and now that it did occur, they stood watching the movements of the bear with feelings of keen apprehension. A shout of joy was heard both from Alexis and Ivan as the great quadruped was seen springing out from the smoke, and clutching to a thick branch that traversed upward near the nest. Embracing the branch with his paws, he commenced descending stern foremost along the limb; but a more rapid descent was in store for him. Out of the four bullets fired into his body, one at least must have reached a mortal part; for his fore arms were seen to relax their hold, his limbs slipped from the bark, and his huge body came "bump" to the ground, where it lay motionless as a log and just as lifeless.

Meanwhile the flames enveloped the nest, and in five minutes more the whole mass was on fire, blazing upward like a beacon. The dry sticks snapped and crackled — the pitchy branches of the pine hissed and spirted — the red cinders shot out like stars, and came showering down to the earth — while high overhead could be heard the vengeful cries of the vultures, as they saw the destruction of their aërial habitation.

But the hunters took no heed of all this. Their task was accomplished, or nearly so. It only remained to divest Bruin of his much-coveted skin; and, having done this in a skilful and proper manner, they mounted their roadsters, and once more took their route across the mountains.

On reaching the first village on the Spanish side, they parted with the expert izzard-hunter and his hired charge — having well remunerated him for his threefold service, each branch of which he had performed to their entire satisfaction.

CHAPTER XXVIII.

SOUTH AMERICAN BEARS.

OUR travellers passed southward to Madrid, where they only remained long enough to witness that exciting but not very gentle spectacle, a bull-fight. Thence proceeding to Lisbon, they took passage direct for Para, or " Gran Para," as it is called — a thriving Brazilian settlement at the mouth of the Amazon river, and destined at no very distant day to become a great city.

The design of our hunters was to ascend the Amazon, and reach, by one of its numerous head-waters, the eastern slope of the Andes mountains — which they knew to be the habitat of the " spectacled bear."

On arriving at Para, they were not only surprised, but delighted, to find that the Amazon river was actually navigated by steamboats ; and that, instead of having to spend six months in ascending to the upper part of this mighty river — as in the olden time — they could now accomplish the journey in less than a score of days ! These steamers are the property of the Brazilian Government, that owns the greater part of the Amazon valley, and that has shown considerable enterprise in developing its resources — much more than any of the Spano-American States, which possess the regions lying

upon the upper tributaries of the Amazon.   It is but **fair**
to state, however, that the Peruvians have also made **an**
attempt to introduce steam upon the Amazon river; and
that they have been unsuccessful, from causes over which
they could scarce be expected **to** have control.   The
chief of these causes appears to have been the dishonesty
of certain American contractors, who provided them **with**
the steamers — three of them — which, on being taken to
the head of steam navigation on the Amazon, were found
to be utterly worthless, and had to be laid up!   This bit
of jobbery is to be regretted the more, since its bad
effects do not alone concern the people of Peru, but the
whole civilized world : for there is not a country on the
globe that would not receive benefit by a development of
the resources of this mighty river.

Our young Russians had been under the belief, as
most people are, that the banks of the Amazon were
entirely without civilized settlements — that the great
river had scarcely been explored — that only a few travellers **had** descended this mighty stream; and that
altogether it was still as much of a *terra incognita* as in
the days of Orellana.   They found that these notions
were quite incorrect ; that not only is there the large
town of Para near the mouth of the Amazon, but there
are other considerable settlements upon its banks, at
different distances from each other, all the way up to
Peru.   Even upon some of its tributaries — as the Rio
Negro and Madeira — there are villages and plantations
of some importance.   Barra, on the former stream, is of
itself a town of 2,000 inhabitants.

In that part of the Amazonian territory which lies

within the boundaries of Brazil, the settlements are, of course, Brazilian — the settlers being a mixture of Portuguese negroes and Christianized Indians. The portion of the great valley higher up towards the Cordilleras of the Andes, belongs to the Spanish-American governments — chiefly to Peru. There are also settlements of a missionary character, the population of which consists almost entirely of Indians, who have submitted themselves to the rule of the Spanish priests. Years ago many of these missionary settlements were in a flourishing condition; but at present they are in a complete state of decay.

Our young Russians found, then, that the great South American river was by no means unknown or unexplored — though as yet no great observer has given an account of it. The different travellers who have descended the Amazon, and written books about it, have all been men of slight capacity, and lacking powers of scientific observation; and one cannot help feeling regret that Humboldt did not choose the Amazon, instead of the Orinoco, as the medium of his valuable researches into the cosmography of South America. Such a grand subject was worthy of such a man.

In ascending the Amazon — which our party did by the Brazilian steamer — they were fortunate in finding on board a very intelligent travelling companion; who gave them much information of the great valley and its resources. This man was an old Portuguese trader, who had spent nearly a lifetime in navigating not only the Amazon itself, but many of its larger tributaries. His business was to collect from the different Indian tribes

the indigenous products of the forests — or *montana*, as
it is called — which stretches almost without interruption
from the Andes to t e Atlantic. In this vast tropical
forest there are many productions that have found their
way into the channels of commerce; and many others
yet unknown or unregarded. The principal articles obtained by the traders are sarsaparilla, Peruvian bark,
annatto, and other dyes, vanilla, Brazil-nuts, Tonka
beans, hammocks, palm fibre, and several other kinds of
spontaneous vegetable productions. Monkeys, toucans,
macaws, parrots, and other beautiful birds, also enter into
the list of Amazonian exports; while the imports consist
of such manufactured articles as may tempt the cupidity
of the savage, or the weapons necessary to him either
in war or for the chase.

In this trade their travelling companion had spent
thirty years of his life; and being a man of intelligence
he had not only acquired a considerable fortune, but
laid in a stock of geographical knowledge, of which the
young Russians were not slow to take advantage. In
the natural history of the *montana* he was well versed;
and knew the different animals and their habits from
actual observation — for which thirty years of adventure
had given him a splendid opportunity. It was a rich
store, and our travellers, especially the naturalist Alexis,
did not fail to draw largely from it.

From the information given by this intelligent trader,
Alexis was enabled to determine several facts about the
bears of South America that had hitherto been doubtful.
He learnt that there are at least two very distinct varieties of them — one, the "spectacled bear" (*ursus ornatus*)

— so called, on account of the whitish rings around his eyes, suggesting the idea of spectacles; and another without these white eye markings, and which has been lately named by a distinguished German naturalist *ursus frugilegus*.

The former kind is known throughout the Peruvian countries as the " Hucumari," and although it inhabits the Cordilleras, it does not ascend to the very cold elevations known as the " paramos" and " puna." On the contrary, it affects a warmer climate, and is not unfrequently found straying into the cultivated valleys termed generally the " Sierra." The *ursus frugilegus* chiefly frequents the tangled woods that cover the eastern spurs of the Andes, ranging often as far down as the montaña, and never so high as the declivities that border on the region of snow.

Both of these species are black bears, and termed "oso negro" by the Spanish-Americans; but the Hucumari is distinguished by a white list under the throat, a white breast, a muzzle of a grayish buff color, and the crescent-like eye markings already mentioned. It is also of a gentler disposition than its congener, smaller in size, and never preys upon other animals. The other does so — frequently making havoc among the flocks of sheep, and even attacking the cattle and horses of the *haciendas*. The *ursus frugilegus* will give battle even to man himself — when baited, or rendered furious by being chased.

Both these species are supposed to be confined to the Chilian and Peruvian Andes. This is an erroneous supposition. They are equally common in Bolivia, and in the sierras of New Grenada and Venezuela. They are

K

found on both sides of Lake Maracaibo — in the sierras Perija and Merida. One of them, at least, has also been observed in the mountains of Guiana — though naturalists have not met with it there. Humboldt, it is true, saw the tracks of what the natives told him **was** *a bear* on the Upper Orinoco ; and, reasoning from their size, **he drew** the inference that it must have been a much smaller species than the *ursus americanus ;* but in this matter the great philosopher was led into an error by a misapplied name. He was informed that the animal was the " oso carnero," or flesh-eating bear — a title given by the Mission Indians to distinguish **it** from two other animals, which they also erroneously term bears — the " oso palmero," or great ant-eater (*tamanoir*), and the " oso hormiguero " (*tamandua*). The animal by whose **tracks** Humboldt was misled, was, no doubt, one of the smaller plantigrade animals (*coatis* or *grisons*), of which there are several species in the forests of South America.

Our hunters **learnt** enough from their travelling acquaintance **to convince them that,** in whatever latitude **they** might approach the Andes from the east, they would be certain to find both varieties of the South American black bear ; but that the best route they could take would be up the great Napo river, which rises not very far from the old Peruvian capital of Quito. In the wild provinces of Quixos and Macas, lying to the east of Quito — **and to** which the Napo **river** would conduct them — they would be certain to meet with the animals they were in search of.

They would have been equally sure of meeting bears in the territory of Jean de Bracamoros ; and this would have been more easily reached; but Alexis knew that by taking that route across the Cordilleras, they would be thrown too far to the west for the Isthmus of Panama — which it was necessary they should cross on their way to the northern division of the American continent.

By keeping up the Napo to its source, and then crossing the Cordilleras of New Granada, they would still be enabled to make westerly as far as Panama — to which port they could get passage in one of the Grenadian coasting-vessels.

On arriving at the mouth of the Napo, therefore, they engaged a *periagua*, with its Indian crew, and continued their journey up this stream towards the still distant Cordilleras of Quito.

CHAPTER XXIX.

THE AMAZONIAN FOREST.

THE river Napo is one of the largest of the head-waters of the Amazon, and one of the most interesting — since, by it, most of the early expeditionists descended in search of the country of the gilded kings, and the gold-roofed temples of Manoa. Though these proved to be fabulous, yet the existence of gold-dust among the Indians of the Napo was true enough, and is true to the present hour. On this river, and its numerous branches, gold-washings, or *placers*, are quite common; and occasionally the savages, who roam over this region, collect the dust, and exchange it with the traders who venture among them. The Indians, however, are of too idle a habit to follow this industry with any degree of energy; and whenever they have obtained a quill full of the metallic sand — just enough to purchase them some coveted nick-nack of civilized manufacture — they leave off work, and the precious ingots are permitted to sleep undiscovered in their beds.

Notwithstanding the length of their journey up the Napo, our travellers did not deem it tedious. The lovely tropical scenery ever under their eyes, together with the numerous little incidents which were constantly occurring, relieved the monotony of their daily life, and kept them in a constant state of interested excitement. At every bend of the river appeared some object, new and worthy of admiration — some grand tropical plant or tree, some strange quadruped, or some bird of glorious plumage.

The craft in which they travelled was that in general use on the upper tributaries of the Amazon: a large canoe — hollowed out from the gigantic *bombax ceiba*, or silk-cotton tree — and usually known as a *periagua*. Over the stern part, or quarter-deck, a little "round-house" is erected, resembling the tilt of a wagon; but, instead of ash-hoops and canvas, it is constructed of bamboos and leaves of trees. The leaves form a thatch to shade the sun from the little cabin inside, and they are generally the large leaves of the *vihai*, a species of *heliconia*, which grows abundantly in the tropical forests of South America. Leaves of the *musacæ* (*plantains and bananas*) serve for a similar purpose; and both kinds are equally employed in thatching the huts in which the natives dwell.

The little cabin thus constructed is called a *toldo*. Inside it is high enough for a man to sit upright, though not to stand; and generally it is only used for sleeping in, or as a shelter during rain. At other times the traveller prefers the open air; and sits or reclines upon the roof of the toldo, which is constructed of sufficient strength to bear his weight. The forward part of the periagua is left quite open; and here the rowers take their stations, so that their movements do not interfere with the comfort of the travellers.

Through the influence of the Portuguese trader, our

party had the good fortune to obtain a proper periagua and crew. They were Christianized Indians, belonging to one of the Spanish missions situated far up the Napo. They had descended this river with a cargo of the products of the mission; and were just about starting to go back, as our travellers arrived at the river's **mouth.** An agreement was easily entered into with the *capataz,* **or** chief of the periagua; and as our travellers always paid liberally for such service, and kept the crew well fed, they received as good attendance and accommodation as circumstances would admit of. **Here** and there on the banks of the river — though at very long intervals apart — were settlements of the wild Indians of the forest; and as nearly all the tribes of Amazonia do less or more in the way of cultivation and commerce, our travellers were enabled from time to time to replenish their larder. Their guns, too, helped materially to keep up the supply: since almost every day game of one kind or another was procured along the banks. For bread they had *farinha,* **a** good stock of which they had brought with them on the steamer from Para. This is the grated root of the manioc plant (*iatropha manihot*), and forms the staple food of all classes throughout the countries of Amazonia.

Alexis was particularly interested in what they **saw.** Never had naturalist a finer field for observation. **Here** was nature presented to the eye in its most normal condition. Here could be observed the tropical forest in all its primeval virginity, unbroken by the axe of the lumberer, and in many places untrodden even by the foot of the hunter. Here its denizens — quadrupeds, *quadru-*

mana, birds, reptiles, and insects — might be seen following out their various habits of life, obedient only to the passions or instincts that had been implanted in them by Nature herself, but little modified by the presence of man. Now would appear a flock of *capivaras* — or *chiguires,* as they are also called — the largest of rodent animals, basking upon some sunny bank, raising their great rabbit-like heads, and gazing curiously at the passing *periagua.* Perhaps before the travellers had lost sight of them, the whole gang would be seen suddenly starting from their attitudes of repose, and in desperate rush making for the water. Behind them would appear the yellow-spotted body of the jaguar — the true tyrant of the Amazonian forest, who, with a single blow of his powerful paw would stretch a *chiguire* upon the grass, and then, couching over his fallen victim, would tear its body to pieces, drink its warm blood, and devour its flesh at his leisure.

If by good fortune the flock might all escape, and reach the water, the jaguar, conscious of their superior adroitness in that element, would at once abandon the pursuit; and returning to his ambush, lie waiting for a fresh opportunity. But for all that, the poor chiguires would not be certain of safety; for even in the water they might encounter another enemy, equally formidable and cruel, in the gigantic *jacare* — the crocodile of the Amazonian waters. Thus assailed in either element, the poor innocent rodents are driven from land to water, and from the water back again to the land; and so kept in a state of continual fear and trembling. The puma, too, assails them, and the *jaguarundi,* and the fierce *coati-mundi;* and not unfrequently the enormous *anaconda*

infolds them in its deadly embrace; for the innocuous
creatures can make no defence against their numerous
enemies; and but for that fecundity which characterizes
the family to which they belong — the so-called "Guinea-
pigs" — their race would be in danger of total extir-
pation.

The chiguires were not the only gregarious animals
observed by our travellers in their ascent of the Napo.
Others of a very different order appeared in the *peccaries*,
or wild pigs of the *montana*. These are true pachyderms,
and in reality pigs; though naturalists have seen fit to
separate **them** from the genus *Sus*, and constitute for
them a genus of their own. It is hardly necessary to
say that this is a very useless proceeding — since the
peccaries are neither more nor less than true wild hogs,
the indigenous representatives of the *suidæ*, on the
American continent. Their classification into a separate
genus has been productive of no good purpose, but the
very contrary: since it has added to the number of
zoölogical names, thereby rendering still more difficult
the study of that interesting science. For such an endless vocabulary, we are chiefly indebted to the specula-
tions of anatomic naturalists, who, lacking opportunities
of actual observation, endeavor to make up for it by
guesses and conjectures, founded upon some little tubercle
upon a tooth! Notwithstanding their learned treatises
it often proves — and very often too — that these
tubercles tell most abominable stories; in plainer terms,
that the animals "lie in their teeth."

The peccary — which the old writers were content to
regard as a wild pig, and very properly placed under the

genus *sus* — is now termed *dicotyles*. Two species only
are yet known to naturalists — the "white-lipped" and
"collared" (*D. labritus* and *collaris*); and although they
are rarely found frequenting the same district of country,
either one or the other kind can be encountered in all
the wilder parts of America — from California on the
north, to the latitude of the La Plata on the south.
Both are nearly of one form and color — a sort of
speckled grayish-brown; the collared species being so
named from a whitish list running up in front of its
shoulders, and forming the semblance of a collar; while
the white-lipped derives its specific title from having lips
of a grayish-white color. In size, however, there is a
great difference between the two: the white-lipped
peccary weighing 100 lbs., or nearly twice the weight of
the collared species. The former, too, is proportionably
stouter in build, and altogether a stronger and fiercer
animal; for although fierceness is not a characteristic of
their nature, like other animals of the hog family, when
roused, they exhibit a ferocity and fearlessness equalling
that of the true *carnivora*.

Both kinds of peccary are preyed upon by the jaguar;
but this tyrant of the wilds approaches them with more
caution and far less confidence, than when he makes his
onslaught on the helpless chiguires; and not unfrequently in conflicts with the peccary, the jaguar comes off
only second best.

Of this fact our travellers had ample proofs — having frequently witnessed, while ascending the Napo, encounters between the peccaries and the jaguars. One
of these encounters they had watched with an interest

more than common : for in its result their own safety was
concerned ; and the very position of peril in which they
were placed, enabled them to have a full and perfect
view of the whole spectacle ; an account of which we
find recorded in the journal of Alexis.

CHAPTER XXX.

THE PERUVIAN CINNAMON-TREE.

They had reached a district which lies between two great branches of the Napo river, and which bears the name of *Canelos,* or the "cinnamon country." The name was given to it by the Spanish discoverers of Peru — from the fact of their finding trees in this region, the bark of which bears a considerable resemblance to the celebrated spice of the East Indies. *Canela* is the Spanish name for cinnamon; and the rude adventurers Pineda and Gonzalez Pizarro, fancying it was the real cinnamon-tree itself, so called it; and the district in which they found it most abundant thenceforward took the name of Canelos.

The tree, afterwards identified and described by the Spanish botanist Mutis, is not the *Laurus cinnamomum* of Ceylon; but a species of *laurus* peculiar to the American continent — to which this botanist has given the name *laurus cinnamomöides.* It is not, however, confined to the region around the Rio Napo, but grows in many parts of the Great *Montana,* as well as in other countries of tropical America. Bonpland identified it on the Upper Orinoco, and again in the county of Caraccas; though nowhere does it appear to be in such plenty as to

the east of the Cordilleras of Ecuador **and Peru** —
throughout the provinces of Quixos, Macas, and Jaen de
Bracamoros.   In these provinces it is found forming
extensive woods, and filling the air with the delicious
aroma of its flowers.   The bark of the *laurus cinnamomöides* **is** not considered equal in delicate flavor to that
of the Oriental cinnamon.   It is hotter and more pun-
gent **to the** taste — otherwise the resemblance between
the two trees **is very** considerable, their foliage being
much alike, and the bark peeling off of nearly equal
thickness.   **The** American, however, becomes more
brownish when dried ; and, **though it is not** equal to
the cinnamon-bark of Ceylon, large quantities of it are
collected, both for use in the Spanish-American countries
and for export to Europe — where it is often passed off
for the true cinnamon.   Were it not that the province of
Canelos is rather inaccessible to commerce, no doubt **a**
great deal more of it would find its way into the European markets ; but there are perils and hardships in the
collecting of this bark, which make it unprofitable to deal
in, even **at the** full **price of the true** cinnamon.   The
Peruvians believe that, were the tree cultivated in a
proper manner, as the Oriental cinnamon is, its bark
would prove equal in quality to the latter ; and perhaps
this may be true, since occasionally specimens of it have
been procured, having all the rich aroma of the spice **of**
Ceylon.   These have been taken from trees that grew in
favorable situations — that is, standing alone, and where
the sun had free access to the leaves and flowers.   The
leaves themselves have the peculiar cinnamon flavor, and
the flowers also ; but in a much stronger degree.   Indeed,

the flowers are even more aromatic than those of the
laurus cinnamomum.

It is said that the wild pigs (*peccaries*) are very fond
of these flowers, as well as the seeds, when ripe; and a
singular habit of these animals is related by some of the
early Peruvian travellers — the Jesuit Ovalle for one.
The old father states that when a flock of the peccaries
go in search of the flowers of the canela-tree, they separate into two divisions, of about nearly equal numbers.
The individuals of one division place their shoulders to
the different trees; and, by shaking them violently, cause
the flowers to fall down to the earth. While thus employed, the peccaries of the other party stand under the
shower, and eat undisturbedly until they have quite filled
their bellies, or otherwise satisfied themselves. These
last then take the place of the hungry hogs; and recip-
rocating the service by shaking the trees, leave the former to enjoy themselves in their turn!

It is not easy to swallow this story of the Jesuit, though
he was himself a native of the country where the scene
is laid. That part of it which relates to the hogs shaking the trees for one another, is not likely to be true,
though it is possible all the other particulars are correct.

It may be true enough that the animals shake the trees
to bring down the flowers: for this would exhibit a
sagacity not greater than hogs of other species are capable of; but it is not according to the laws of their moral
nature to perform the service for one another. That
they roam in great flocks through the canela forests, and
devour with avidity the blossoms of these trees, is undoubtedly a fact — of which our travellers had the evi-

dence of their own eyes while on their journey up the river Napo.

They were passing a place where these wild cinnamon-trees lined the banks of the stream; and, in order to make a closer examination of such an interesting species, Alexis landed from the periagua. Ivan went along with him — taking his double-barrelled gun, in hopes of getting a shot at something. In one barrel he had a bullet, while the other was loaded with shot — so that he was prepared for any sort of game that might turn up, either beasts or birds. Alexis, as usual, carried his rifle.

It was their intention to walk for some distance up the bank. There was a sandy strip between the water and the trees — which would enable them to make way without difficulty — and it is only where this occurs that the banks of the Amazonian rivers can be followed on foot. Generally, the thick forest comes down to the very water's edge; and there is no pathway except an occasional track followed by the chiguires, tapirs, and other animals; but, as these creatures only open the underwood to the height of their own bodies, all above that is a matted labyrinth of leaves and llanos, that form an impenetrable barrier to the passage of anything so tall as a man. The Indians themselves rarely follow these paths, but keep to their canoes or periaguas.

Seeing this fine open sand-bar, which appeared to stretch for miles above them, our young travellers, tired of sitting upon the *toldo*, determined to stretch their legs in a walk; and, directing the capataz to keep up the river and take them in above, they set out along the bank — now and then dipping into the woods, wherever an opening showed itself, and examining such rare natural objects as attracted their attention.

Pouchskin did not go with them; and the reason was that, some days before, Pouchskin had encountered a mishap, by which he was laid up lame. The cause of his lameness was simply that some *chigas* had got between his toes; and not having been extracted in time, had there laid their eggs, and caused a terrible inflammation to his feet, a misfortune that frequently happens in tropical countries. The wound caused by the *chiga*, though not absolutely of fatal consequences, is very dangerous to be trifled with — often leading to the necessity of amputating the part attacked by these diminutive insects. Pouchskin, sneering at the insignificance of the enemy, had neglected taking proper precautions — notwithstanding that the Indian canoe-men had warned him of the danger. The consequence was a swelling of the parts and an inflammation, that lamed the old grenadier as completely as if his leg had been carried off by a bomb-shell; and he was now reclining along the top of the toldo, unable to stand upon his feet.

For this reason, being in no condition to join his young masters on their pedestrian excursions, he was necessarily left behind. It was, perhaps, just as well for him: since it was the means of keeping him clear of a scrape into which both of the young hunters chanced to fall very soon after; and which, perhaps, had Pouchskin been with them, might have ended worse than it did: since it could not have ended much better.

CHAPTER XXXI.

A SKURRY OVER A SAND-BAR.

JOURNEYING along the bank, as we have described, Alexis and Ivan had gone some two or three miles up the river. They were beginning to get tired of their walk: as the sand was rather soft, and sank under their feet at every step. Just then they descried, a little ahead of them, a long bar, or "spit" of the bank, running out nearly to the middle of the river. They made up their minds to go on until they should reach this bar. At its end appeared a proper place for the periagua to come to, and take them aboard.

The craft was still working up-stream, and had got nearly opposite them, so that they could hail it. They did so — desiring the *popero*, or steersman, to put in at the extremity of the sand-bar. This matter having been arranged, they continued on up the bank, going at their leisure.

On arriving at that part of the bank where the sand-spit projected into the river, they were about stepping out upon it, when the quick ear of Ivan caught the sound of some animals moving among the underwood. All was game that came to Ivan's gun; and as he had seen nothing worth wasting a charge upon, during their long

walk, he was very desirous to have a shot at something
before returning to the periagua.

What he heard was a rustling of leaves. It did not
appear to proceed from any particular spot, but rather
from all parts of the forest. Now and then the sound
was varied by a sort of half-squeaking half-grunting
noise, that indicated the presence of animals, and a great
many of them too: since, at times, several scores of
these squeaks and grunts could be heard uttered simultaneously. Alexis heard the sounds too ; but being less
of a keen sportsman than his brother, cared less to go
after the creatures that were making them. He had no
objection to Ivan straying a little out of his way ; and
promised to wait for him on the open bank.

Had he known what sort of game it was that his
brother was going after — that is, had he been acquainted
with the habits of the animals that were making themselves heard, he would either have gone along with Ivan,
or, what is more likely, would have hindered him from
going at all. Alexis, however, was under the impression
that monkeys of some kind were making the strange
noises — for not only are there many species of these in
the forests of the Napo, but some that can imitate the
voices of other animals. Of course, with monkeys, there
could be no danger : since none of the American quadrumana are large enough or strong enough to attempt
an attack upon man.

Ivan had not left the spot more than five minutes, when
a loud report, reverberating among the trees, announced
that he had fired his gun ; and, almost in the same instant,
a second crack told that both barrels were now empty.

8 *

L

Alexis was about proceeding to the place to see what his brother had shot, when all at once his ears were assailed by a loud chorus of noises — a screaming, and snorting, and grunting — that seemed to come from all parts of the wood; while the cracking of sticks, and the "swishing" of branches, announced a singular commotion — as if some hundreds of creatures were rushing to and fro through the jungle. At the same instant was heard the voice of Ivan, crying out in accents of alarm; while the boy was himself seen breaking his way through the bushes, and running with all his might in the direction of his brother. His looks betokened terror, as if some dreaded pursuer was behind him.

"Run! brother — run!" cried he, as he got clear of the underwood; "run for your life! — they're after me — they're after me!"

It was no time to inquire what pursuers were after him. Evidently, they were of a sort to be shunned: since they had caused to the courageous Ivan such serious alarm; and Alexis, without staying for an explanation, turned, and joined in his brother's flight. Both directed themselves towards the open sand-spit, in hopes of being able to reach the periagua — which could be seen just drawing up to its point of the bar.

They had not made a dozen steps into the open ground, when the bushes from which they had just parted were seen to vibrate, and from out their trembling cover rushed a host of strange creatures: literally a host, for, in a few seconds' time, not less than two hundred of them made their appearance.

They were quadrupeds of a grayish-brown color, not

larger than half-grown pigs; and pigs they were — that is to say, they were *peccaries*. They were those of the species *labiatus* — as could be seen by their white lips. These lips were especially conspicuous, for each individual was rushing on open-mouthed, with snout raised aloft — all of them cracking their teeth like castanets, uttering, as they ran, a confused chorus of short, sharp grunts and squeaks, expressive of anger.

As soon as Alexis saw them, he recognized the peril of the situation in which he and his brother were placed. He had read, and heard moreover from the Portuguese trader — as well as from the Indian canoe-men — of the danger to be apprehended from an attack of these fierce little animals; and how the hunter, to escape from them, is often compelled to take to a tree. Had he and Ivan reflected for a moment, they would probably have made for the woods, instead of running out on the open sand-bar, as they had done. It was now too late, however. The peccaries covered the whole line of beach behind them; and no tree could have been reached, without passing back again through the midst of the drove. Their retreat in the direction of the woods was completely cut off; and there appeared no alternative, but to make the best use they could of their heels, and, if possible, get on board the periagua.

With this determination they rushed on over the sand-bank, closely pursued by the peccaries.

CHAPTER XXXII.

PURSUED BY PECCARIES.

I<small>T</small> is needless to say that our young hunters took as
long strides as the nature of the ground would permit ;
but, unfortunately, they were not long enough. The sand
was soft and heavy, and in places so full of holes, where
the turtles had had their eggs — now empty — that the
fugitives could make but slow progress, though fear was
urging them to do their utmost. The pursuers themselves did not make as good speed as they would have
made on firmer ground, but they were going faster than
the pursued ; and the boys were beginning to fear that
they would never be able to reach the periagua in time.
To be overtaken meant the same as to be dragged upon
the sand, and to be torn to pieces by the sharp tusks
of the peccaries. The periagua was still three hundred
yards distant. The Indians saw the chase, and knew the
danger — knew it so well, that it was not likely they
would venture ashore to the rescue ; and as for Pouchskin, he **was** unable to budge an inch — even had there
been no other means of saving his young masters. It
was a moment of fearful apprehension for the faithful
Pouchskin. He had seized his fusil, and wriggled his
body into an erect attitude ; but he felt powerless to do
more.

In this moment of peril an object came under the eyes of Alexis that promised safety. At least it held out the prospect of a temporary retreat from the danger — though whether they might succeed in reaching this retreat was not certain.

This object was a tree — not standing and growing, but a fallen tree — dead, and divested of its leaves, its bark, and most of its branches. It lay upon the sand-spit — where it had, no doubt, been deposited during the season of floods — not exactly in the line of their flight, but some paces to the right of the track they would have followed in keeping on to the periagua. It was nearer them than the boat, by full two hundred yards; and Alexis observing this, suddenly conceived a hope that they might yet reach the tree, and find shelter, either upon its trunk or among its branches. Of these the larger ones still remained — rising many feet above the surface of the sand, and shrouded under masses of weeds and withered grass, which had been there deposited at the falling of the flood. Indeed, Alexis scarce looked to the capabilities the tree afforded for giving them a secure retreat. There was no alternative. It was like the drowning man catching at straws. He only cast a look behind him, to see what time they might have to spare; and by a quick glance calculating their distance from the pursuers, he shouted to Ivan to follow him, and turned obliquely towards the tree.

They had noticed the tree when first starting to run, but had not thought of it as a place of retreat. Indeed, they had thought of nothing except getting back to the boat; and it was only now, when this had proved clearly impossible, that they determined on taking to the tree.

As they faced full towards it, they were able to note the chances it offered for their safety. They saw that they were not so bad; and, encouraged by hope, they made efforts more energetic than ever — both of them straining every nerve and muscle in their legs and bodies.

The effort was needed; but fortunately it proved sufficient to save them. Just sufficient: for scarce had they succeeded in getting upon the log, and drawing their limbs up after them, when the infuriated host arrived upon the ground, and in a few seconds surrounded them on all sides. Lucky it was that the log was a large one. It was the dead-wood of a gigantic silk-cotton — the *bombax ceiba* of the tropical forests; and its trunk, being full five feet in diameter, gave them that elevation above the surface of the sand.

Notwithstanding this, they saw that their safety was not yet quite assured: for the spiteful peccaries, instead of desisting in their attacks, commenced leaping up against the log, endeavoring to reach its top, and there assail them. Now and then one more active than the rest actually succeeded in getting its fore feet over the ridge of the dead-wood: and, had it not been for the quick use which our hunters made of the buts of their guns, undoubtedly they would have been reached. Both stood with their barrels grasped firmly — now threatening the assailing host, and now punching in the head such of them as sprang within reach — the peccaries all the while uttering their angry grunts, and chattering their teeth, as if a hundred strings of Christmas crackers were being let off at the same time!

In this way the conflict was carried on — the hunters, bit by bit, working themselves along the log towards the top branches, which, projecting higher, appeared to offer a more secure place of retreat. But at intervals as they advanced, they were compelled to make halt, and deal a fresh shower of blows to their assailants, who still kept leaping up from below.

At length the boys succeeded in reaching the projecting limbs of the tree; and each choosing one strong enough to carry him, they scrambled up towards their tops. This placed them in a position where they could set the peccaries at defiance; for although the creatures could now spring up on the main trunk — which several of them had already done — the more slender limbs baffled all their efforts at climbing; and such of them as attempted it were seen to roll off and tumble back upon the sand-bank.

The hunters, now feeling secure, could not refrain from a shout of joy, which was answered by a cheer from the periagua, in which the baritone of Pouchskin bore a conspicuous part.

Our heroes now believing themselves in for a siege, began to consider the best means of raising it; when all at once a spectacle came under their eyes, that guided their thoughts into a far different channel.

CHAPTER XXXIII.

SCYLLA AND CHARYBDIS.

THEIR retreat **upward upon** the slanting limbs of the tree had brought a large band of their assailants round **to that side; and,** just as **they raised** their triumphant cry, they saw **the** peccaries dancing among the branches that lay extended along the sand-bar. Many of these were hidden by the flakes of hanging grass already mentioned; but another fearful **creature** chanced to have been hidden there also; who now displayed himself in all his shining majesty — not only **to** the **eyes of the** besieged, but likewise to those of the besiegers. **The creature was** a quadruped — one of fearful mien, and dimensions **far** exceeding that of the Lilliputian peccaries. **It was** their **natural** enemy — the jaguar!

Whether **it** was the shout that **had** startled him, or the peccaries had trodden him out of his lair, or both, certain it was that he now sprang suddenly out, and with one bound launched himself upon the log For a moment he stood cowering **on** its top, turning his eyes first upon the branches where the boys had taken refuge, and **then** in the opposite direction, towards the woods. He seemed irresolute **as to** which course he would take; and this irresolution, so long as it lasted, produced an unpleasant effect upon our young hunters. Should the jaguar also attack them, their destruction might be accounted as certain; for the great cat would either strike them down from their unstable perch, or claw them to death if they continued to cling to it. Of course, to fall down among the peccaries would be death, equally certain and terrible.

By good fortune, however, the jaguar at the moment of showing himself was eagerly assailed by the wild pigs; and it was to escape from their assault, that he had sprung upward to the log. Thither the peccaries had pursued him, and were now endeavoring to reach the top of the dead-wood, just as they had done while after the hunters. The jaguar no longer stood silent and irresolute; but, uttering loud screams, he commenced defending himself against the assailing host, striking them with his broad ungulated paws, and flinging one after another back to the ground, where they lay kicking in the throes of death.

Perhaps it was the presence of mind exhibited by Alexis that brought matters to a climax, and saved the lives of himself and his brother. His rifle was still loaded — for it had appeared useless firing into the midst of two hundred assailants. He knew he could kill *only one or two;* and this, instead of frightening them off, would but render the others more implacable in their resentment. Partly for this reason, and partly that he had all along held the piece "clubbed" in his hands, he had reserved his fire. Now was the time to deliver it. The jaguar was even more to be dreaded than the pec-

caries — for they were now secure from the attacks of
the latter, whereas they were not only within reach of
the former, but in the very place to which the brute
might fancy retreating. To prevent this contingency,
Alexis resolved to give the jaguar his bullet.

It was but a moment's work to turn the gun in his
hand and take aim. The crack followed quickly; and,
on the instant, the hunters had the gratification to see
the great tawny quadruped spring out from the log, and
alight upon the sand — where, in a second's time, he was
surrounded by the dark drove, that from all sides rushed
screaming towards him.

It was a bit of good fortune that the bullet of Alexis
had only wounded the jaguar, instead of killing him on
the spot. Had he been shot dead, the peccaries would
have torn his beautiful skin to ribbons, and reduced his
quivering flesh to mince-meat, and that within the space
of a score of seconds; but luckily it chanced that the
jaguar was only wounded — had only received a broken
leg; and, availing himself of the three that remained
sound, he commenced retreating towards the timber.
Thither he was followed by his thick-skinned assailants;
who, transferring their spite to this new enemy, seemed
to forget all about their original adversaries, who remained quietly perched upon the limbs of the tree!

For some time nothing could be seen but a confused
crowd, writhing over the sand — a dark mass, in the
midst of which now and then a bright yellow object appeared conspicuous, and was then for a time out of sight;
and thus, like a rolling wave, the great drove went surg-
ing on, amidst grunting and screaming, and growling, and

chattering of teeth, till it swept up to the edge of the underwood, and then suddenly disappeared from the eyes of the spectators!

Whether the peccaries eventually succeeded in destroying the jaguar, or whether the wounded tyrant of the forest escaped from their terrible teeth, could never be told. Our young hunters had no curiosity to follow and witness the *dénouement* of this strange encounter. Neither cared they to take up the bodies of the slain. Ivan was completely cured of any *penchant* he might have had for peccary pork; and, as soon as their late assailants were fairly out of sight, both leaped down from the limbs of the tree, and made all haste towards the boat. This they reached without further molestation; and the canoe-men, rapidly plying their paddles, soon shot the craft out upon the bosom of the broad river — where they were safe from the attack either of wild pigs or wild cats.

It was likely the jaguar betook himself to a tree — his usual mode of escape when surrounded by a herd of infuriated peccaries — and, as a proof that he had done so, our travellers could hear the wild hogs still uttering their fierce grunts long after the boat had rounded the sandspit, and was passing up the bend of the river.

CHAPTER XXXIV.

THE OLD MISSIONS.

PASSING many scenes of interest, and meeting with several other strange incidents, our travellers at length arrived at Archidona — a small town at the head of boat navigation upon the Napo, and the usual port of embarkation for persons proceeding from the country around Quito to the regions upon the Amazon. Up to this place they had been journeying through a complete wilderness — the only exceptions being some missionary stations, in each of which a monkish priest holds a sort of control over two or three hundred half-christianized Indians. It would be absurd to call these missions civilized settlements: since they are in no degree more advanced, either in civilization or prosperity, than the *maloccas,* or villages of the wild Indians — the "infidels," as it pleases the monks to call those tribes who have not submitted to their puerile teachings. Whatever difference exists between the two kinds of Indians is decidedly in favor of the unconverted tribes, who display at least the virtues of valor and a love of liberty, while the poor neophytes of the missions have suffered a positive debasement, by their conversion to this so-called " Christian religion." All these monkish settlements — not only on the Napo, but on the other tributaries of the Amazon — were at one time in a state of considerable prosperity. The missionary padres, backed by a little soldier help from the Spanish Government, were more able to control their Indian converts, and compel them to work — so that a certain amount of prosperity was visible in the mission settlements, and some of them had even attained to a degree of wealth. This, however, was but an apparent civilization; and its benefits only extended to the monks themselves. The Indian neophytes were in no way bettered by the wealth they created. Their condition was one of pure slavery — the monks being their masters, and very often hard taskmasters they proved themselves — living in fine conventual style upon the sweat and labor of their brown-skinned converts. The only return made by them to the Indians was to teach the latter those trades, by the practice of which they themselves might be benefited, and that was their sole motive for civilizing them. On the other hand, instead of endeavoring to cultivate their intellectual nature, they strove in every way to restrain it — inculcating those doctrines of duty and obedience so popular among the priests and princes of the world. They taught them a religion of the lips and not of the heart — a religion of mere idle ceremonies, of the most showy kind; and above all a religion whose every observance required to be paid for by toll and tithe. In this manner they continued to filch from the poor aboriginal every hour of his work — and keep him to all intents and purposes an abject slave. No wonder, that when the Spanish power declined, and the soldier could no longer be spared to secure the authority of the

priest — no wonder that the whole system gave way, and
the missions of Spanish America — from California to
the Patagonian plains — sank into decay. Hundreds of
these establishments have been altogether abandoned —
their pseudo-converts having returned once more to the
savage state — and the ruins of convents and churches
alone remain to attest that they ever existed. Those still
in existence exhibit the mere remnants of their former
prosperity, and are only kept together by the exertions
of the monks themselves — backed by a slight thread of
authority, which they derive from the superstitions they
have been able to inculcate. In fact, in the missions now
existing, the monks have no other power than that which
they wield through the terrors of the Church; and in
most cases, these *padres* constitute a sort of hierarch
chieftaincy, which has supplanted the old system of **the**
curacas, or caciques.

At one period the missions of the Napo were both
numerous **and** powerful. That was while they were
under the superintendence of those active apostles, the
Jesuit fathers; but **most of** their settlements have long
ago disappeared; and now only a few sparse stations
exist along the **borders** of the great *Montaña*.

In ascending the Napo, our travellers had an opportunity of visiting some of these old missionary establishments; and observing the odd rigmarole of superstitions
there practised under the guise, and in the name of religion — a queer commingling of pagan rites with Christian ceremonies — not unlike those Budhistic forms from
which these same ceremonies have been borrowed.

One advantage our travellers derived from the existence of these stations: they were enabled to obtain from them the provisions required upon their long riverine voyage; and without this assistance they would have found it much more difficult to accomplish such a journey.

Beyond Archidona the rest of the journey to Quito would have to be performed on horseback, or rather mule-back; but they were not going direct to Quito. Between them and the old Peruvian capital lay the eastern cordillera of the Andes, and it was along its declivities, and in the valleys between its transverse spurs, facing the Montaña, they would have to search for the haunts of the bear.

On the Napo itself, still higher up than Archidona — where the stream, fed by the snows of the grand volcano of Cotopaxi, issues from the spurs of the Andes — there were they most likely to accomplish the object of their expedition, and thither determined they to go.

Having procured mules and a guide, they proceeded onward; and after a journey of three days — in which, from the difficulty of the roads, they had travelled less than fifty miles — they found themselves among the foot-hills of the Andes — the giant Cotopaxi with his snowy cone towering stupendous above their heads.

Here they were in the proper range of the bears — a part of the country famous for the great numbers of these animals — and it only remained for them to fix their head-quarters in some village, and make arrangements for prosecuting the chase.

The little town of Napo, called after the river, and situated as it is in the midst of a forest wilderness, offered all the advantages they required; and, choosing it as their temporary residence, they were soon engaged in searching for the black bear of the Cordilleras.

CHAPTER XXXV.

EATING A NEGRO'S HEAD.

ACCORDING to their usual practice, they had hired
one of the native hunters of the district to act as a guide,
and assist them in finding the haunts of Bruin. In Napo
they were fortunate in meeting with the very man, in
the person of a *mestizo*, or half-blood Indian, who followed hunting for his sole calling. He was what is
termed a "tigrero," or tiger-hunter — which title he
derived from the fact that the jaguar was the principal
object of his pursuit. Among all Spanish Americans —
Mexicans included — the beautiful spotted jaguar is erroneously termed *tigre* (tiger), as the puma or couguar is
called *leon* (lion). A hunter of the jaguar is therefore
denominated a " tiger-hunter," or *tigrero*.

There are no puma or lion-hunters by profession —
as there is nothing about this brute to make it worth
while — but hunting the jaguar is, in many parts of
Spanish America, a specific calling ; and men make their
living solely by following this occupation. One inducement is to obtain the skin, which, in common with those
of the great spotted cats of the Old World, is an article
of commerce, and from its superior beauty commands
a good price. But the *tigrero* could scarce make out to

live upon the sale of the skins alone; for although a London furrier will charge from two to three guineas for a jaguar's robe, the poor hunter in his remote wilderness market can obtain little more than a tenth part of this price, — notwithstanding that he has to risk his life, before he can strip the fair mantle from the shoulders of its original wearer.

It is evident, therefore, that jaguar-hunting would not pay, if there was only the pelt to depend upon; but the *tigrero* looks to another source of profit — the *bounty*.

In the hotter regions of Spanish America, — the Brazils as well — there are many settlements to which the jaguar is not only a pest, but a terror. Cattle in hundreds are destroyed by these great predatory animals; even full-grown horses are killed and dragged away by them! But is this all? Are the people themselves left un-molested? No. On the contrary, great numbers of human beings every year fall victims to the rapacity of the jaguars. Settlements attempted on the edge of the great Montaña — in the very country where our young hunters had now arrived — have, after a time, been abandoned from this cause alone. It is a well-known fact that where a settlement has been formed, the jaguars soon become more plentiful in that neighborhood: the increased facility of obtaining food — by preying on the cattle of the settlers, or upon the owners themselves — accounting for this augmentation in their numbers. It is precisely the same with the royal tiger of India, as is instanced in the history of the modern settlement of Singapore.

To prevent the increase of the jaguars, then, a bounty

9 M

is offered for their destruction. This bounty is sometimes the gift of the government of the country, and
sometimes of the municipal authorities of the district.
Not unfrequently private individuals, who own large
herds of cattle, give a bounty out of their private purses
for every jaguar killed within the limits of their estates.
Indeed, it is not an uncommon thing for the wealthy proprietor of a cattle-estate (*hacienda de ganados*) to maintain one or more " tigreros " in his service — just as
gamekeepers are kept by European grandees — whose
sole business consists in hunting and destroying the jaguar. These men are sometimes pure Indians, but, as a
general thing, they are of the mixed, or *mestizo* race. **It**
need hardly be said that they are hunters of the greatest
courage. They require to be so : since an encounter
with a full-grown jaguar is but little less dangerous than
with his striped congener of the Indian jungles. In
these conflicts, the tigreros often receive severe wounds
from the teeth **and** claws of their terrible adversary ;
and, not unfrequently, the hunter himself becomes the
victim.

You may wonder that men are found to follow such a
perilous calling, and with such slight inducement — for
even the bounty is only a trifle of a dollar or two — differing in amount in different districts, and according to
the liberality of the bestower. But it is in this matter as
with all others of a like kind — where the very danger
itself seems to be the lure.

The tigrero usually depends upon fire-arms for destroying his noble game ; but where his shot fails, and it is
necessary to come to close quarters, he will even attack

the jaguar with his *machete* — a species of half-knife
half-sword, to be found in every Spanish-American cot-
tage from California to Chili.

Very often the jaguar is hunted without the gun. The
tigrero, in this case, arms himself with a short spear, the
shaft of which is made of a strong, hard wood, either a
guaiacum, or a piece of the split trunk of one of the
hard-wood palms.

The point of this spear is frequently without iron —
only sharpened and hardened by being held in the fire
— and with this in his left hand, and his short sword in
the right, the hunter advances with confidence upon his
formidable adversary. This confidence has been forti-
fied by a contrivance which he has had the precaution to
adopt — that is, of enveloping his left arm in the ample
folds of his blanket — *serape, roana,* or *poncho,* according
to the country to which he belongs — and using this as
a shield.

The left arm is held well forward, so that the woollen
mass may cover his body against the bound of the animal, and thus is the attack received. The jaguar, like
all feline quadrupeds, springs directly forward upon his
prey. The tigrero, prepared for this, and, with every
nerve braced, receives the assailant upon the point of his
short spear. Should the jaguar strike with its claws it
only clutches the woollen cloth; and while tearing at
this — which it believes to be the body of its intended
victim — the right arm of the hunter is left free, and
with the sharp blade of his *machete* he can either make
cut or thrust at his pleasure. It is not always that the
tigrero succeeds in destroying his enemy without receiv-

ing a scratch or two in return; but a daring hunter makes light of such wounds — for these scars become badges of distinction, and give him *éclat* among the villages of the Montaña.

Just such a man was the guide whom **our** young hunters had engaged, and who, though a tiger-hunter **by profession**, was equally expert at the capturing **of a bear** — when one **of these** animals chanced to stray down from the higher slopes of the mountains, into the warmer country frequented by the jaguars. It was not always **that** bears could **be** found in these lower regions; but there is a particular season of the year when the black bear (*ursus frugilegus*) descends far below his usual range, and even wanders far out into the forests of the Montaña.

Of course there must be some inducement for his making this annual migration from his mountain home; for the *ursus frugilegus*, though here dwelling within the tropics, does not affect a tropical climate. Neither is he **a** denizen of the very cold plains — the *paramos* — that extend among the summits of eternal snow. A medium temperature **is** his choice; and this, as we have already stated, he finds among the foot-hills forming the lower zone of the Eastern Andes. It is there he spends most **of** his life, and that is his place of birth, and consequently his true home. At a particular season of the year, corresponding to the summer of our own country, he makes a roving expedition to the lower regions; and for what purpose? This was the very question which Alexis put to the tigrero. The answer was as curious as laconic :—

"*Comer la cabeza del negro.*" (To eat the negro's head !)

"Ha, ha! to eat the negro's head!" repeated Ivan, with an incredulous laugh.

"Just so, señorito!" rejoined the man; "that is what what brings him down here."

"Why, the voracious brute!" said Ivan; "you don't mean to say that he makes food of the heads of the poor negroes?"

"O no!" replied the tigrero, smiling in his turn; "it is not that."

"What then?" impatiently inquired Ivan. "I've heard of negro-head tobacco. He's not a tobacco-chewer, is he?"

"*Carrambo!* no, señorito," replied the tiger-hunter, now laughing outright; "that's not the sort of food the fellow is fond of. You'll see it presently. By good luck, it's just in season now — just as the bears fancy it — or else we need n't look to start them here. We should have to go further up the mountains: where they are more difficult both to find and follow. But no doubt we'll soon stir one up, when we get among the *cabezas del negro.* The nuts are just now full of their sweet milky paste, of which the bears are so fond, and about a mile from here there are whole acres of the trees. I warrant we find a bear among them."

Though still puzzled with this half-explanation, our young hunters followed the guide — confident that they would soon come in sight of the "negro's head."

CHAPTER XXXVI.

THE TAGUA-TREE.

AFTER going about a mile further, as their guide had forewarned them, they came within sight of a level valley, or rather a plain, covered with a singular vegetation. It looked as if it had been a forest of palms — the trunks of which had sunk down into the earth, and left only the heads, with their great radiating fronds above the ground! Some of them stood a foot or two above the surface; but most appeared as if their stems had been completely buried! They were growing all the same, however; and, at the bottom of each great bunch of pinnate leaves, could be seen a number of large, roundish objects — which were evidently the fruits of the plant.

There was no mystery about the stems being buried underground. There were no stems, and never had been any — except those that were seen rising a yard or so above the surface. Neither was there any longer a mystery about the "negro's head;" for the rounded fruit with its wrinkled coriaceous pericarp — suggesting a resemblance to the little curly knots of wool on the head of an African — was evidently the object to which the tigrero had applied the ambiguous appellation.

What our hunters saw was neither more nor less than

a grove of *Tagua*-trees — better known as the "vegetable ivory."

This singular tree was for a long time regarded as a plant of the *Cycas* family; and by some botanists it has been classed among the *Pandanaceæ*, or screw-pines. Growing, as its leaves do, almost out of the earth, or with only a short trunk, it bears a very marked resemblance to the cycads; but for all this, it is a true palm. Its not having a tall trunk is no reason why it should not be a palm, since many other species of *palmaceæ* are equally destitute of a visible stem. It is now, however, acknowledged by the most expert botanists, that the "Tagua" — or "Cabeza del Negro," as the Peruvians style it — is a palm; and it has been honored as the representative of a genus (*Phytelephas*), of which there are but two species known — the great-fruited and little-fruited (*macrocarpa* and *microcarpa*). Both are natives of the hot valleys of the Andes, and differ very little from each other; but it is the species with the larger fruit that is distinguished by the figurative title of "negro's head."

The Peruvian Indians use the pinnate fronds of both species for thatching their huts; but it is the nuts of the larger one that have given its great celebrity to the tree. These are of an oblong triangular shape; and a great number of them are enclosed in the pericarp, already described. When young, they are filled with a watery liquid that has no particular taste; though regarded by the Indians as a most refreshing beverage. A little older, this crystal-like fluid turns of a milky color and consistence; and still later it becomes a white paste.

When fully ripe, it congeals to the whiteness and hard-
ness of ivory itself; and, if kept out of water, is even
more beautiful in texture than the tusks of the elephant.
It has been employed by the Indians from time im-
memorial in the construction of buttons, heads for their
pipes, and many other purposes. Of late years it has
found its way into the hands of civilized artisans; and
since it can be procured at a cheaper rate, and is quite
equal to the real ivory for many useful and ornamental
articles, it has become an important item of commerce.

But however much the vegetable ivory may be esteemed by the Indians, or by bipeds of any kind, there
is one quadruped who thinks quite as much of it as they,
and that is the black bear of the Andes (*ursus frugilegus*).
It is not, however, when it has reached the condition of
ivory that Bruin cares for it. Then the nut would be
too hard, even for his powerful jaws to crack. It is when
it is in the milky state — or rather after it has become
coagulated to a paste — that he relishes it; and with so
much avidity does he devour the sweet pulp, that at this
season he is easily discovered in the midst of his depredations, and will scarce move away from his meal even
upon the appearance of the hunter! While engaged in
devouring his favorite negro-head, he appears indifferent to any danger that may threaten him.

Of this our hunters had proof, and very shortly after
entering among the tagua-trees. As the tigrero had
predicted, they soon came upon the "sign" of a bear,
and almost in the same instant discovered Bruin himself browsing upon the fruit.

The young hunters, and Pouchskin too, were about

getting ready to fire upon him ; when, to their surprise, they saw the tigrero, who was mounted on a prancing little horse, spur out in front of them, and gallop towards the bear. They knew that the killing of the animal should have been left to them ; but as they had given their guide no notice of this, they said nothing, but looked on — leaving the tigrero to manage matters after his own way.

It was evident that he intended to attack the bear, and in a peculiar fashion. They knew this by seeing that he carried a coil of raw-hide rope over his arm, on one end of which there was a ring and loop. They knew, more-over, that this was a celebrated weapon of the South Americans — the *lazo,* in short ; but never having wit-nessed an exhibition of its use, they were curious to do so ; and this also influenced them to keep their places.

In a few minutes the horseman had galloped within some twenty paces of the bear. The latter took the alarm, and commenced trotting off ; but with a sullen reluctance, which showed that he had no great disposi-tion to shun the encounter.

The ground was tolerably clear, the taguas standing far apart, and many of them not rising higher than the bear's back. This gave the spectators an opportunity of witnessing the chase.

It was not a long one. The bear perceiving that the horseman was gaining upon him, turned suddenly in his tracks, and, with an angry growl, rose erect upon his hind legs, and stood facing his pursuer in an attitude of defiance. As the horseman drew near, however, he ap-peared to become cowed, and once more turning tail,

9 *

shambled off through the bushes. This time he only ran
a few lengths: for the shouts of the hunter provoking
him to a fresh fit of fury, caused him to halt again, and
raise himself erect as before.

This was just the opportunity of which the hunter was
in expectation; and before the bear could lower himself
on all-fours — to charge forward upon the horse, the
long rope went spinning through the air, and its noose
was seen settling over the shoulders of the bear. The
huge quadruped, puzzled by this mode of attack, endeavored to seize hold of the rope; but so thin was the
raw-hide thong, that he could not clutch it with his great
unwieldy paws; and by his efforts he only drew the
noose tighter around his neck.

Meanwhile, the hunter, on projecting the lazo, had
wheeled, with the quickness of thought; and, driving his
sharp spurs into the ribs of his horse, caused the latter
to gallop in the opposite direction. One might have supposed that he had taken fright at the bear, and was endeavoring to get out of the way. Not so. His object
was very different. The lazo still formed a link of connection between the hunter and his game. One end of
it was fast to a staple firmly embedded in the wood of the
saddle-tree, while the other, as we have seen, was noosed
around the bear. As the horse stretched off, the rope
was seen to tighten with a sudden jerk; and Bruin was
not only floored from his erect attitude, but plucked clear
off his feet, and laid sprawling along the earth. In that
position he was not permitted to remain: for the horse
continuing his gallop, he was dragged along the ground
at the end of the lazo — his huge body now bounding

several feet from the earth, and now breaking through the bushes with a crackling, crashing noise, such as he had himself never made in his most impetuous charges.

In this way went horse and bear for half a mile over the plain ; the spectators following after to witness the ending of the affair. About that there was nothing particular : for when the tigrero at length halted, and the party got up to the ground, they saw only an immobile mass of shaggy hair — so coated with dust as to resemble a heap of earth. It was the bear without a particle of breath in his body ; but, lest he might recover it again, the tigrero leaped from his horse, stepped up to the prostrate bear, and buried his *machete* between the ribs of the unconscious animal.

That, he said, was the way they captured bears in his part of the country. They did not employ the same plan with the jaguars : because these animals, crouching, as they do, offered no opportunity for casting the noose over them ; and, besides, the jaguars haunt only among thick woods, where the lazo could not be used to advantage.

Of course, the skin of this particular bear was not suitable for the purpose for which one was required ; and the tigrero kept it for his own profit. But that did not signify : another bear was soon discovered among the tagua-trees ; and this being despatched by a shot from the rifle of Alexis, — supplemented, perhaps, by a bullet from the fusil of the ex-guardsman, — supplied them with a skin according to contract ; and so far as the *ursus frugilegus* was concerned, their bear-hunting in that neighborhood was at an end. To find his cousin with the "goggle eyes," they would have to journey onward and upward ; and adopting for their motto the spirit-stirring symbol, " Excelsior !" they proceeded to climb the stupendous Cordilleras of the Andes.

In one of the higher valleys, known among Peruvians as the " Sierra," they obtained a specimen of the " Hucumari." They chanced upon this creature while he was engaged in plundering a field of Indian corn — quite close to a "tambo," or traveller's shed, where they had put up for the night. It was very early in the morning when the corn-stealer was discovered ; but being caught in the act, and his whole attention taken up with the sweet milky ears of maize, his " spectacled " eyes did not avail him. Our hunters, approaching with due caution, were able to get so near, that the first shot tumbled him over among the stalks.

Having secured his skin, they mounted their mules, and by the great Cordillera road proceeded onward to the ancient capital of northern Peru.

CHAPTER XXXVII.

NORTHWARD !

AFTER resting some days in the old capital of Quito, our travellers proceeded to the small port of Barbacoas, on the west coast of Equador ; and thence took passage for Panama. Crossing the famous Isthmus to Porto Bello, they shipped again for New Orleans, on the Mississippi. Of course, their next aim was to procure the North American bears — including the Polar, which is equally an inhabitant of northern Asia, but which, by the conditions of their route, would be more conveniently reached on the continent of North America. Alexis knew that the black bear (*ursus americanus*) might be met with anywhere on that continent from the shores of Hudson's Bay to the Isthmus of Panama, and from the seaboard of the Atlantic to the coast of the Pacific Ocean. No other has so wide a range as this species — with the exception, perhaps, of the brown bear of Europe — which, as we have said, is also an Asiatic animal. Throughout the whole extent of country above defined, the black bear may be encountered, not specially confining himself to mountain-ranges. True, in the more settled districts he has been driven to these — as affording him a refuge from the hunter; but in his normal condition

he is by no means a mountain-dwelling animal. On the
contrary, he affects equally the low-wooded bottoms of ra-
vines, and is as much at home in a climate of tropical or
sub-tropical character, as in the cold forests of the Canadas.

Mr. Spencer Baird — the naturalist intrusted by the
American Government to describe the *fauna* of their
territory, and furnished for his text with one of the
most splendid collections ever made — in speaking of
the genus *ursus*, makes the following remarks : —

"The species of bears are not numerous, nor are
they to be found except in the temperate regions of the
northern hemisphere. **North** America possesses more
species than any other part of the world, having at
least four, and perhaps five."

With the exception of the very idle assertion that
"the species of bears are not numerous," every idea
put forth in the above categorical declaration is the very
reverse of what is true.

Is the polar bear found only in the temperate regions
of the northern hemisphere? Is the *ursus arctos* of
Europe confined to these limits? Are the bears of
South America? — the sloth-bear of India and Ceylon?
— the bruang of Borneo? — and his near congener,
the bruang of Java and Sumatra? Why, these last
are actually dwellers among palm-trees — as the cocoa-
planters know to their cost! Even Mr. Baird's own
American black bear is not so "temperate" in his habits; but loves the half-tropical climate of Florida and
Texas quite as much as the cold declivities of the
Alleghanies.

And how does North America possess more species

than any other part of the world? Even admitting the doubtful fifth, on the continent of Asia, there are six species at the very least; and, if we are allowed to include the Oriental islands, we make eight Asiatic. There are three species in the Himalaya mountains alone — unquestionably distinct, dwelling in separate zones of altitude, but with the territory of all three visible at a single *coup d'œil.*

Mr. Baird is a naturalist of great celebrity in America. He is a secretary of the Smithsonian Institution: he should make better use of the books which its fine library can afford him.

The United States Government is extremely unfortunate in the selection of its scientific *employés* — more especially in the departments of natural history. Perhaps the most liberal appropriation ever made for ethnological purposes — that for collecting a complete account of the North American Indians — has been spent without purpose, the "job" having fallen into the hands of a "placeman," or "old hunker," as the Americans term it — a man neither learned nor intellectual. With the exception of the statistics furnished by Indian agents, the voluminous work of Schoolcraft is absolutely worthless; and students of ethnology cannot contemplate such a misappropriation without feelings of regret.

Fortunately, the American aboriginal had already found a true portrayer and historian. Private enterprise, as is not unfrequently the case, has outstripped Government patronage in the performance of its task. In the unpretending volumes of George Catlin we find the most complete ethnological monograph ever given to

the world; but just for that reason, Catlin, not Schoolcraft, should have been chosen for the "job."

Knowing the range of the black bear to be thus
grandly extended, our young hunters had a choice of
places in which to look for one; but, as there is no
place where these animals are more common than in
Louisiana itself, they concluded that they could not do
better than there choose their hunting-ground. In the
great forests, which still cover a large portion of Louisiana, and especially upon the banks of the sluggish
bayous, where the marshy soil and the huge cypress-
trees, festooned with Spanish moss, bid defiance to all
attempts at cultivation, the black bear still roams at will.
There he is found in sufficient numbers to insure the
procuring of a specimen without much difficulty.

The hunters of these parts have various modes of
capturing him. The log-trap is a common plan; but
the planters enjoy the sport of running him down with
dogs; or rather should it be termed running him up;
since the chase usually ends by Bruin taking to a tree
and thus unconsciously putting himself within reach of
the unerring rifle.

It was by this means that our young hunters determined to try their luck; and they had no difficulty in
procuring the necessary adjuncts to insure success. The
great Czar, powerful everywhere, was not without his
agent at New Orleans. From him a letter of introduc-
tion was obtained to a planter living on one of the interior *bayous;* and our heroes, having repaired thither,
were at once set in train for the sport — the planter
placing himself, his house, his hounds, and his horses at
their disposal.

CHAPTER XXXVIII.

THE NORTHERN FORESTS.

On their arrival, the hospitable planter sent to his neighbors, and arranged a grand hunt, to come off at an early day, specified in the invitation. Each was to bring with him such hounds as he was possessed of — and in this way a large pack might be got together, so that a wide extent of forest could be driven.

Among the planters of the Southern States this is a very common practice : only a few of them keeping what might be called a regular kennel of hounds, but many of them having five or six couples. In a neighborhood favorable to the chase, by uniting a number of these little bands together, a pack may be got up large enough for any purpose.

The usual game hunted in the Southern States is the American fallow-deer (*cervus virginianus*), which is still found in considerable plenty in the more solitary tracts of forest all over the United States. It is the only species of deer indigenous to Louisiana : since the noble stag or "elk," as he is erroneously called (*cervus canadensis*), does not range so far to the south. On the Pacific coast this animal is found in much lower latitudes than on that of the Atlantic.

N

Besides the fallow-deer, the fox gives sport to the Louisiana hunter. This is the gray fox *(vulpes virginianus)*. The bay lynx also — or wild-cat, as it is called *(lynx rufus)* — and now and then, but more rarely, the cougar *(felis concolor)*, give the hounds a run before taking to the tree.

Raccoons, opossums, and skunks are common enough in the forests of Louisiana; but these are regarded as "vermin," and are not permitted to lead the dogs astray.

With regard to the other animals mentioned, they all rank as noble game — especially the cougar, called "panther" by the backwoodsman — and the pack may follow whichever is first "scared up."

The grand game, however, is the bear; and the capture of Bruin is not a feat of every-day occurrence. To find his haunts it is necessary to make an excursion into the more unfrequented and inaccessible solitudes of the forest — in places often many miles from a settlement. Not unfrequently, however, the old gentleman wanders abroad from his unknown retreat, and seeks the plantations — where in the night-time he skulks round the edges of the fields, and commits serious depredations on the young maize plants, or the succulent stalks of the sugar-cane, of which he is immoderately fond. Like his brown congener of Europe he has a sweet tooth, and is greatly given to honey. To get at it he climbs the bee-trees, and robs the hive of its stores. In all these respects he is like the brown bear; but otherwise he differs greatly from the latter species, so much indeed, that it is matter of surprise how any naturalist should have been led to regard them as the same.

Not only in color, but in shape and other respects, are they totally unlike. While the fur of the brown bear is tossed and tufty — having that appearance usually termed *shaggy* — that of the American black bear is of uniform length, and all lying, or rather standing, in one direction, presenting a smooth surface corresponding to the contour of his body. In this respect he is far more akin to the bears of the Asiatic islands, than to the *ursus arctos*. In shape, too, he differs essentially from the latter. His body is more slender, his muzzle longer and sharper, and his profile is a curve with its convexity upward. This last characteristic, which is constant, proclaims him indubitably a distinct species from the brown bear of Europe ; and he is altogether a smaller and more mild-tempered animal.

As the grand "chasse" had been arranged to come off on the third day after their arrival, our young hunters determined to employ the interval in ranging the neighboring woods ; not with any expectation of finding a bear — as their host did not believe there was any so near — but rather for the purpose of becoming acquainted with the character of the North American *sylva*.

That of South America, Alexis had carefully observed and studied in their long journey across that continent. He had noted the grand tropical trees — the palms and *pothos* plants — the *mimosas* and *musaceæ* — the magnificent forms of the *bombax* and *bertholletia* — the curious *cecropias* and fig-trees — the giant *cedrelas* and the gum-yielding *siphonias*. On the Andes he had observed the agaves, the cycads, and cactaceæ — all strange to the eye of a Russian. He was now desirous of making

himself familiar with the forests of North America;
which, though of a sub-tropical character in Louisiana,
contained forms altogether different from those of the
Amazonian regions.   Here he would meet with the
famed magnolia, and its relative the tulip-tree; the ca-
talpa and flowering cornel, the giant cypress and sycamore, the evergreen oak, the water-loving tupelo, and
the curious fan-like palmetto.   Of these, and many other
beautiful trees belonging to the North American *sylva*,
Alexis had read — in fact, knew them botanically; but
he wished to cultivate a still pleasanter acquaintance
with them, by visiting them in their own native home.

For this purpose he and Ivan set out alone, with only
a negro for their guide; the planter being engaged, visiting his different friends, and warning them for the grand
hunt.

Pouchskin remained behind.   He had been left at the
house — to do some necessary repairs to the travelling
traps both of himself and his young masters, which, after
their long South American expedition, needed looking
to.   At this work had Pouchskin been left, surrounded
by a circle of grinning darkies, in whose company the
old grenadier would find material to interest and amuse
him.

It was only for a stroll that our young hunters had
sallied forth, and without any design of entering upon
the chase; but they had become so accustomed to carrying their guns everywhere, that these were taken along
with them.   Some curious bird or quadruped might be
started — whose fur or feathers they might fancy to
make an examination of.   For that reason, both shoul-
dered their guns.

CHAPTER XXXIX.

THE LONE LAGOON.

THEY were soon beyond the bounds of the plantation, and walking under the dark majestic woods — the darkey guiding them on their way. They had heard of a curious lake or lagoon, that lay about a mile from the plantation. There they would be likely to witness a spectacle characteristic of the swamps of Louisiana; and thither they directed their steps.

Sure enough, on arriving at the borders of the lagoon, a singular scene was presented to their eyes. The whole surface of the lake appeared alive with various forms of birds and reptiles. Hundreds of alligators were seen, lying like dead trees upon the water, their corrugated backs appearing above the surface. Most of them, however, were in motion, swimming to and fro, or darting rapidly from point to point, as if in pursuit of prey. Now and then their huge tails could be seen curling high up in air, and then striking down upon the water, causing a concussion that echoed far through the forest. At intervals a shining object, flung upward by their tails, could be seen for a moment in the air, amidst the showery spray that was raised along with it. It was easy to see that the glittering forms thus projected were fishes,

214 BRUIN.

and that it was the pursuit of these that was causing the commotion among the huge reptiles. Aquatic birds, of a great number of kinds, were equally busy in the pursuit of the fish. Huge pelicans stood **up** to their tibia in the water — now and then immersing their long mandibles, and tossing their finny victims high into the air. Cranes and herons too were there — among **others the** tall Louisiana crane — conspicuous among the smaller species — snow-white egrets, **the** wood ibis, and others of white and roseate hue — the snake-darter, with long-pointed beak and crouching serpent-like neck — the qua-bird, **of** lugubrious note and melancholy aspect — and, fairest of **all, the** scarlet flamingo.

Other birds besides those **of** aquatic habits took part in the odd spectacle. Hovering in the air were black vultures — the carrion-crow and the turkey-buzzard — and upon the tops of tall dead trees could be seen **the** king of the feathered multitude, the great white-headed eagle. His congener, the osprey, soared craftily above — at intervals swooping down, and striking his talons into the fish which the alligators **had** tossed into the air — thus robbing the reptiles of their prey, to be robbed in turn by his watchful cousin-german upon the tree.

The spectacle was far from being a silent one: on the contrary, the confused chorus of sounds was deafening to the ears of the spectators. The hoarse bellowing of the alligators — the concussions made by their great tails striking the water — the croaking of the pelicans, and the clattering of their huge mandibles — the doleful screaming of the herons, cranes, and qua-birds — the shrieks of the osprey — and the shrill maniac laughter of the white-headed eagle, piercing through all other sounds — formed a medley of voices as unearthly as inharmonious.

A shot from the gun of Ivan, that brought down a splendid specimen of the white-headed eagle — together with the appearance of the hunters by the edge of the water — put a sudden termination to this grand drama of the wilderness. The birds flew up into the air, and went soaring off in different directions over the tops of the tall trees; while the huge reptiles, that had been taught by the alligator-hunters to fear the presence of man, desisted for a while from their predatory prey, and retreated to the reeds upon the opposite shore.

The spectacle was one well worthy of being seen, and one that cannot be witnessed every day — even in the swamps of Louisiana. Its occurrence at that time was accounted for by the drying up of the lake, which left the fish at the mercy of their numerous enemies.

Having taken up the eagle which Ivan had shot, the young hunters continued their excursion along the edge of the lagoon.

They had not gone far when they came upon a bank of mud, that had formerly been covered with water. So recently had the water dried from it, that, in spite of the hot sun shining down upon it, the mud was still soft. They had not gone many steps further, when they perceived upon its surface, what at first they supposed to be the tracks of a man. On getting a little closer, however, they doubted this; and, now recollecting the resemblance which they had noticed in the snows of Lapland — between the footsteps of a human being and those of a bear

— it occurred to them that these might also be bear-
tracks — though they knew that the tracks of the American bear would be slightly different from those of his
European cousin.

To satisfy themselves, **they** hastened forward **to** examine the tracks; but their negro guide had anticipated
them, and now called out, with the whites of his eyes
considerably enlarged, —

"Golly, young mass'rs! dat be de tracks ob um ba!"

"**A** bear!"

"Ya, ya, mass'rs! a big ba — dis child know um track
— **see'd um** many de time — de **ole coon** he be arter de
fish too — all ob **dem adoin' a bit ob fishin'** dis mornin'
— yaw, yaw, yaw!"

And the darkey laughed at what he appeared **to consider** an excellent joke.

On closely scrutinizing the tracks, Alexis and Ivan
saw that they were in reality the tracks of a bear —
though much **smaller** than those they had followed in
Lapland. They **were** quite fresh — in fact, so recently
did they appear **to have been made, that** both at the same
time, and by an involuntary impulse, raised their eyes
from the ground and glanced around them; as if they
expected to see the bear himself.

No such animal was in sight, however. **It was** quite
probable he had been on the ground, at their first coming
up to the lake; but the report of Ivan's gun had alarmed
him, and he had made off into the woods. This was
quite probable.

"What a pity," reflected Ivan, "that I did n't leave
the eagle alone! We might have got a sight of Master

Bruin, and given him the shot instead. And now," added he, "what's to be done? There's no snow,— therefore we can't track the brute. The mud-bank ends here, and he's gone off it, the way he came. Of course he wouldn't be out yonder among those logs. He wouldn't have taken shelter there, would he?"

As Ivan spoke, he pointed to a little peninsula that jutted out into the lake, some twenty or thirty yards beyond the spot where they were standing. It was joined to the mainland by a narrow neck or isthmus of mud; but at the end towards the water there was a space of several yards covered with dead trees — that had been floated thither in the floods, and now lay high and dry, piled irregularly upon one another.

Alexis looked in the direction of this pile as Ivan pointed it out.

"I'm not so sure of that," he answered, after scrutinizing the logs. "It's a likely enough place for an animal to lurk. He might be there."

"Let us go and see, then!" said Ivan, "If he's there he can't escape us, without our having a shot at him; and you say that these American bears are much easier killed than ours. The South Americans were so, certainly. I hope their northern brothers may die as easy."

"Not all," rejoined Alexis. "We may expect some tough struggles when we come to the great grisly, and to him of the polar regions; but the black bears are, as you conjecture, not so difficult to deal with. If wounded, however, they will show fight; and though their teeth and claws are less dangerous than the others, they can

give a man a most uncomfortable hug, I have heard.
But let us go, as you say. If not yonder, he must have
taken to the woods. In that case there is no way of
following him up, except by dogs ; **and for** these we must
go back to the house."

As they continued talking, they advanced towards the
narrow isthmus that connected the little peninsula **with**
the mainland.

"What a pity," remarked Ivan, "that that great log
is there ! But for it we might have seen his track in the
mud crossing over."

Ivan referred to a prostrate trunk that traversed the
isthmus longitudinally — extending from the mainland to
the higher ground of the peninsula, to which it formed a
kind of bridge or causeway. Certainly, had it not been
there, either the bear's tracks would have been **seen in**
the mud or not; and if not, then no bear could have
passed over to the peninsula, and their exploration would
have **been** unnecessary. But, although they saw no
tracks, they had started to examine the wood-pile ; and
they continued on, climbing up to the log, and walking
along its top.

All at once, Alexis was seen to pause and bend his
body forward and downward.

"What is it?" inquired Ivan, who was behind, on
seeing his brother in the bent attitude, as if he looked at
something on the log.

"The bear's tracks !" answered Alexis, in a low but
earnest tone.

"Ha! you think so? Where?"

Alexis pointed to the dead-wood under his eyes —

upon the bark of which were visible, not the tracks of a bear, but dabs of mud, that must have been recently deposited there, either by the feet of a bear or some other animal.

"By the Great Peter!" said Ivan, speaking cautiously, notwithstanding his innocent adjuration; "that must be his tracks? It's the same sort of mud as that in which we've just been tracing him — black as ink nearly. It has come off his great paws — not a doubt of it, brother?"

"I think it is likely," assented Alexis, at the same time that both looked to the locks of their guns, and saw that the caps were on the nipples.

A little further along the log, the bark was smoother, and there the track was still more conspicuous. The print was better defined, and answered well for the footmark of a bear. There was the naked paw, and the balls of the five toes, all complete. They no longer doubted that it was the track of a bear.

It was just a question whether the animal had gone over the log and returned again. But this was set at rest, or nearly so, by a closer scrutiny. There was no sign of a return track. True, he might have washed his paws in the interval, or cleaned them on the dead-wood; but that was scarce probable, and our hunters did not think so. They felt perfectly sure that the bear was before them; and, acting upon this belief, they cocked their guns, and continued their approach towards the wood-pile.

CHAPTER XL.

A DARKEY ON BEAR-BACK.

Both the young hunters succeeded in passing over the log, **and** had set foot on the peninsula; while the negro, who was following a little behind, was still upon the prostrate trunk. Just at that moment a noise was heard — very similar to that made by a pig when suddenly started from its bed of straw — a sort of half snort, half grunt; and along with the noise a huge black body was seen springing up from under the loose pile of dead trees, causing several of them to shake and rattle under its weight. Our hunters saw at a glance that it was the bear; and levelled their guns upon it with the intention of firing.

The animal had reared itself on its hind legs — as if to reconnoitre the ground — and while in this attitude both the hunters had sighted it, and were on the eve of pulling their triggers. Before they could do so, however, the bear dropped back on all-fours. So sudden was the movement, that the aim of both was quite disconcerted, and they both lowered their guns to get a fresh one. The delay, however, proved fatal to their intention. Before either had got a satisfactory sight upon **the** body of the bear, the latter sprang forward with a fierce growl, and rushed

right between the two, so near that it was impossible
for either of them to fire otherwise than at random. Ivan
did fire, but to no purpose ; for his bullet went quite wide
of the bear, striking the log behind it, and causing the
bark to splinter out in all directions. The bear made no
attempt to charge towards them, but rushed straight on
— evidently with no other design than to make his escape to the woods. Alexis wheeled round to fire after
him ; but, as he was raising his gun, his eye fell upon
the negro, who was coming on over the log, and who had
just got about half-way across it. The bear had by this
time leaped up on the other end, and in a hurried gallop
— that had been quickened by the report of Ivan's
piece — was going right in the opposite direction. The
negro, who saw the huge shaggy quadruped coming
straight towards him, at once set up a loud " hulla-ballo,"
and, with his eyes almost starting from their sockets, was
endeavoring to retreat backwards, and get out of the way.

His efforts proved fruitless : for before he had made
three steps to the rear, the bear — more frightened at the
two adversaries behind him than the one in front —
rushed right on, and in the next instant pushed his
snout, head, and neck between the darkey's legs!

Long before this the negro had lost his senses, but
now came the loss of his legs : for as the thick body of
the bear passed between them, both were lifted clear up
from the log, and hung dangling in the air. For several
feet along the log was the negro carried upon the bear's
back, his face turned to the tail ; and no doubt, had he
preserved his equilibrium, he might have continued his
ride for some distance further. But as the darkey had

no desire for such a feat of equestrianism, he kept struggling to clear himself from his involuntary mount. His body was at length thrown heavily to one side, and its weight acting like a lever upon the bear, caused the latter to lose his balance, and tumbling off the log, both man and bear fell " slap-dash " into the mud.

For a moment there was a confused scrambling and spattering, and splashing, through the soft mire — a growling on the part of the bear, and the wildest screeching from the throat of the affrighted negro — all of which came to an end by Bruin — whose body was now bedaubed all over with black mud — once more regaining his feet, and shuffling off up the bank, as fast as his legs could carry him.

Alexis now fired, and hit the bear behind; but the shot, so far from staying his flight, only quickened his pace; and before the darkey had got to his feet, the shaggy brute had loped off among the trees, and disappeared from the sight of everybody upon the ground.

The grotesque appearance of the negro, as he rose out of the mire in which he had been wallowing, coated all over with black mud — which was a shade lighter than his natural hue — was too ludicrous for Ivan to resist laughing at; and even the more serious Alexis was compelled to give way to mirth. So overcome were both, that it was some minutes before they thought of reloading their guns, and giving chase to the bear.

After a time, however, they charged again; and crossing back over the log, proceeded in the direction in which Bruin had made his retreat.

They had no idea of being able to follow him without

dogs; and it was their intention to send for one or two to the house, when they perceived that the bear's trace could be made out — at least, for some distance — without them. The inky water, that had copiously saturated his long fur, had been constantly dripping as he trotted onward in his flight; and this could easily be seen upon the herbage over which he had passed.

They determined, therefore, to follow this trail as far as they could; and when it should give out, it would be time enough to send for the dogs.

They had not proceeded more than a hundred yards; when all at once the trail trended up to the bottom of a big tree. They might have examined the ground further, but there was no need; for, on looking up to the trunk, they perceived large blotches of mud, and several scratches upon the bark, evidently made by the claws of a bear. These scratches were, most of them, of old date; but there were one or two of them quite freshly done; besides, the wet mud was of itself sufficient proof that the bear had gone up the tree, and must still be somewhere in its top. The tree was a sycamore, and therefore only sparsely covered with leaves; but from its branches hung long festoons of Spanish moss (*tillandsia usneoides*), that grew in large bunches in the forks — in several of which it was possible even for a bear to have stowed himself away in concealment.

After going round the tree, however, and viewing it from all sides, our hunters perceived that the bear was not anywhere among the moss; but must have taken refuge in a hollow in the trunk — the mouth of which could be seen only from one particular place; since it was hidden on all other sides by two great limbs that led out from it, and between which the cavity had been formed by the decaying of the heart-wood.

There could be no doubt that Bruin had entered this tree-cave; for all around the aperture the bark was scraped and worn; and the wet mud, lately deposited there, was visible from below.

CHAPTER XLI.

CUTTING OUT THE BEAR.

The question was, how he was to be got out? Perhaps by making a noise he might issue forth?

This plan was at once tried, but without success. While the negro rasped the bark with a pole, and struck the stick at intervals against the trunk, the hunters stood, with guns cocked, watching the hole, and ready to give the bear a reception, the moment he should show himself outside.

It was all to no purpose. Bruin was too cunning for them, and did not protrude even the tip of his snout out of his secure cavity.

After continuing the rasping, and repeating the blows, till the woods echoed the sonorous concussions, they became convinced that this plan would not serve their purpose, and desisted from it.

On examining the track more closely, they now perceived spots of blood mixed among the mud which the bear had rubbed off upon the bark. This convinced them that the animal was wounded, and therefore there would be no chance of starting him out from his hole. It was no doubt the wound that had led him to retreat to this tree, so near the place where he had been attacked, otherwise he would have led them a longer chase through the woods before attempting to hide himself. When severely wounded, the black bear betakes himself to the first hollow log or tree he can find; and taking refuge in it, will there remain — even to die in his den, if the wound has been a fatal one.

Knowing this habit of the animal, our hunters perceived that they had no chance of again setting their eyes upon the bear, except by cutting down the tree; and they resolved to adopt this method, and fell the great sycamore to the ground.

The darkey was despatched to the plantation; and soon returned with half a dozen of his brethren, armed with axes — Pouchskin heading the sable band. Without further delay the chopping began; and the white chips flew out from the great trunk in all directions.

In about an hour's time the sycamore came crashing down, carrying a number of smaller trees along with it. The hunters, who expected that the bear would at once spring forth, had taken their position to cover the mouth of the cavity with their guns; but, to their surprise, the tree fell, and lay as it had fallen, without any signs of Bruin. This was strange enough; for, as the negroes alleged, in all similar cases the bear is certain to charge out upon the fall of a tree that contains him!

A sapling was now obtained, and inserted into the cavity — at first with caution, but after a time it was punched in with all the force that Pouchskin could put into his arm. He could feel the bear quite distinctly; but poke the animal as he might, it would not stir.

It was suggested that they should cut into the trunk

— at a place opposite to where the bear was encased — and then they could drag him out at will ; and, although this would cost a good deal of trouble, it appeared to be the only mode of reaching the obstinate animal.

This course was followed, therefore ; and a cross section being made of the hollow trunk, the shaggy hair was at length reached and then the body of Bruin, who was found to be dead as a nail !

They no longer wondered that he had paid no heed to the punching of the pole. The bullet of Alexis had traversed his huge body in a longitudinal direction, until it had lodged in a vital part, and, of course, it was this that had deprived him of life. He would, therefore, have died all the same, and in his tree-den, too, whether they had pursued him or not.

Our hunters learnt from their negro assistants a singular fact in relation to the black bear : and that is, that the tree-cavity in which the animal often takes shelter, or goes to sleep, is rarely of greater width than his own body ! In most cases it is so narrow, that he cannot turn round in it, nor has it any lair at the bottom wide enough for him to lie down upon. It follows, therefore, that he must sleep in a standing position, or squatted upon his hams. It is in this attitude he makes his descent into the cavity, and in the same way comes down the trunk of the tree, when at any time making his departure from his den. From this it would appear that the upright attitude is as natural to this animal, as that of resting on all-fours, or even lying prostrate on the ground ; for it is well known that, farther to the north — where the winters are more severe, and where the black bear hy-

bernates for a short season — he often takes his nap in a
tree-cavity, which his body completely fills, without the
possibility of his turning round in it! One precaution
he takes, and that is to scrape off all the rotten wood
around the sides of the cavity ; but for what purpose he
exercises this curious instinct, neither hunter nor naturalist can tell. Perhaps it is that the projections may not
press against his body, and thus render his couch uncomfortable ?

Our young hunters found this bear one of the largest
of his species, and his skin, after the mud had been
washed off, proved to be an excellent specimen.

Of course, they coveted no other ; but for all that,
they had the pleasure of being present at the death of
several bears, killed in the great hunt that came off on
the appointed day.

A deer-chase was also got up for their special entertainment — during which a cougar was "treed" and
killed — an event of rarer occurrence than even the death
of a bear ; for the cougar is now one of the scarcest quadrupeds to be met with in the forests of North America.

Another entertainment which the planter provided
for his guests was a "barbecue" — a species of festival
peculiar to the backwoods of America, and which, on
account of its peculiarity, deserves a word or two of
description.

CHAPTER XLII.

THE SQUATTER'S BARGAIN.

As we have just said, the barbecue is a festival which especially belongs to the backwoods settlements, although it has now become known even in the older States, and often forms a feature in the great political meetings of an election campaign — losing, however, much of its true character in the elaborate adornments and improvements sometimes bestowed upon it.

When Alexis and Ivan strolled down in the early morning to the quiet glade which had been selected as the scene of this rural festivity, they found there a noisy and bustling crowd. A monstrous fire of logs, enough to roast not only a single ox, but a hecatomb of oxen, was blazing near the edge of the glade, while a half-dozen chattering negroes were busy digging a great pit close by. This pit when entirely excavated, measured some ten or twelve feet in length, by five or six in width, and perhaps three in depth; and was lined with smooth flat stones. As soon as the logs had ceased to flame and smoke, and were fast falling into a mighty heap of glowing ruddy coals, they were shovelled hastily into the pit. Another party of negroes had been busy in the woods, searching out the tall slender saplings of the pawpaw

(*asimina triloba*), and now returned, bringing their spoil
with them.   The saplings were laid across the top of the
pit, thus extemporizing over it a huge gridiron.   The ox,
which was to form the staple of the day's feast, had been
killed and dressed ; and having been split in halves after
the fashion of the barbecue was laid upon the bars to
roast.   Proudly presiding over the operation was the
major-domo of the planter's household, assisted by several
celebrated cooks of the neighborhood, and a score of
chosen farm-hands, whose strength was ever and anon
invoked to turn the beef ; while the *chef* ordered a fresh
basting, or himself sprinkled the browning surface with
the savory dressing of pepper, salt, and fine herbs, for
the composition of which he had attained a grand repu-
tation.

The morning wore swiftly on in the observation of
these novel manœuvres ; and with the noon came the
guests in numbers from the neighboring plantations and
settlements.   Even the determined resistance of the
toughest beef must have failed before the hot attack
of such an army of live coals, as had lain intrenched
in the deep fireplace ; and the tender joints of the
enormous *bœuf roti* were ready to bear their share
in the festivities almost as soon as the invited company.   Separated with great cleavers, and laid into
white buttonwood trays hollowed out for the purpose,
they were borne rapidly to the shady nook selected for
the dining-place, followed by vast supplies of sweet
potatoes, roasted in the ashes, and of rich, golden, maize,
bread.   A barrel of rare cider was broached ; while
good old-fashioned puddings, and the luscious fruits of

the region completed the bill of fare in honor of the day. Of course "joy was unconfined." Everybody pronounced the roast a grand success; and the young Russians thought that they had never tasted so appetizing a meal. With the exhilaration of the fresh, clear air, the encouragement of hearty appetite, and the full flavor of the meat — for it is well known that the sap which exudes from the pawpaw, when thus exposed to fire, adds a new relish to whatever is cooked upon it — combined to make a dinner fit for the Czar himself; and they determined to attempt, at some time, an imitation of the southern barbecue under the colder sky of Russia.

Merriment was unbounded; healths were drunk, songs sung, odd speeches made, and stories told.

One of the last, in particular, made an impression upon our heroes; partly, because it was a bear story, and partly because it illustrated a very characteristic phase of squatter life and practical humor. In fact, Alexis made a sketch of it in his journal, and from his notes we now reconstruct it.

Two squatters had occupied lands not far from each other, and within some eight or ten miles of a small town. Busied in clearing off the woodland, each bethought himself of a source of revenue beyond the produce of his tilled ground. He would occupy an occasional leisure day in hauling to the town, the logs which he cut from time to time, and then selling them as firewood. This unity of purpose naturally brought the two men into competition with one another for the limited custom of the settlement; and a rivalry sprang up be-

tween them, which was fast ripening into jealousy **and**
ill-will, when a curious coincidence occurred.

Each owned a single yoke of oxen, which he used
regularly in his farm labor, and also in dragging his wood
to market. Within a week each lost an **ox**; **one** dying
of some bovine distemper,—the other being so injured
by the fall of a tree, that his owner had been obliged
to kill him.

As one ox could not draw a wood-wagon, the occupation of both squatters as wood-merchants was gone—
and even farm operations were likely to suffer. Each
soon heard of his neighbor's predicament; and proposed
to himself to make a bargain **for the** remaining ox,
that he might be the possessor of the pair, continue
his clearing prosperously, and command the wood-
hauling business. But, as one might suppose, where
both parties were so fully bent upon accomplishing their
own ends, the trade was no nearer a conclusion when a
dozen negotiations had taken place than at first. So
matters stood in *statu quo*, the days rolled by, and our
two squatters found their condition waxing desperate.

One fine morning, squatter the first started off to
make a last attempt—determined to close the bargain
peaceably if he could, forcibly if he must. Revolving
project upon project in his mind, he had traversed the
two or three miles of woodland which lay between him
and his neighbor's clearing, and was just entering it,
when a sudden rustle and significant growl coming from
behind broke in upon his reverie. Turning hastily, he
saw almost at his heels a bear of the most unprepossess-
ing aspect. To reach the cabin before Bruin could over-

take him was impossible; and to turn upon the creature would be folly: for, in the depth of his deliberation, he had forgotten on leaving home to take any kind of weapon with him. Some dead trees had been left standing in the field, and to one of these he sped with flying steps, hoping to find shelter behind it till help could come. He did not hope in vain for this protection. He found that by pretty active dodging he could keep the trunk of the tree between himself and the bear — whose brain could hardly follow the numerous shifts made by the squatter to escape the frequent clutches of his claws. Rising indignantly upon his hind legs, the bear made a fierce rush at the squatter, but hugged only the tough old tree, in whose bark he buried deep his pointed claws. An inspiration flashed through the squatter's mind, as he saw the bear slowly and with some difficulty dragging out his nails; and seizing Bruin's shanks just above the paws, he braced himself against the tree, resolved to try and hold the claws into their woody sockets until his neighbor could respond to his halloos for help.

The other squatter heard his cries; but instead of hastening to the rescue, he came slowly along, carelessly shouldering his axe. Perceiving his neighbor's difficulty, a new solution of the ox question had entered his mind; and to the redoubled appeals for assistance, he calmly replied —

"One one condition, neighbor!"

"What is it?" anxiously inquired the other.

"If I let you loose from the bar, you'll gi' me up your odd steer."

There was no help for it, and with a heavy sigh, the

prisoner consented. " Stop!" cried he, ere the axe could fall; "this old brute has half plagued the life out o' me, and I 'd like nothing better 'n the satisfaction o' killin' him myself. Jest you ketch hold here, and let me give him his death-blow."

The second squatter, rejoicing beyond measure at having accomplished his long-desired purpose, unsuspiciously agreed, dropped the axe, cautiously grasped the sinewy shanks, and bent his strength to the momentary struggle. To his utter dismay, he beheld his neighbor quietly shoulder the axe, and walk away from the ground!

"Hold on!" he shouted; "ain't ye goin' to kill the bar?"

" Wal, not jest now, I fancy; I thought *you* might like to hang on a while?"

The tables thus turned, the deluded squatter had no resource but to make terms with his grimly gleeful neighbor, who at last consented to put an end to the wild beast's life, if he might not only be released from the bargain he had just made, but, in addition, be himself the recipient of the odd ox. Sorely chagrined, the second squatter consented. But he was a little comforted at the idea of a slight *revanche* that had just entered his head. Watching his chance, as the other approached to deal the fatal blow, with a desperate effort he tore out the bear's claws from the bark — setting the infuriated animal free — and then fled at full speed to his cabin, leaving the two original combatants to fight it out between themselves.

The particulars of the contest even tradition has not preserved — the sequel to the narrative only telling that half an hour later the first squatter, scratched and bloody, hobbled slowly up to the cabin, remarking satirically as he threw down the broken axe : —

"Thar, neighbor; I'm afraid I've spiled yer axe, but I'm sure I've spiled the bar. Prehaps you'd let one o' your leetle boys drive that ere ox over to my house ?"

    *   *   *   *   *

After enjoying the hospitality of their planter friend for a few days longer, our travellers once more resumed their journey; and proceeded up the great Mississippi, towards the cold countries of the North.

CHAPTER XLIII.

THE POLAR BEAR.

A FEW weeks after leaving the Louisiana planter, our hunters were receiving hospitality from a very different kind of host, a "fur-trader." **Their** head-quarters was Fort Churchill, **on** the western **shore** of Hudson's Bay, and once the chief entrepôt of the famous company **who** have so long directed the destinies of that extensive region — sometimes styled Prince Rupert's Land, but more generally known as the "Hudson's Bay Territory."

To Fort Churchill they had travelled almost due **north** — first up **the** Mississippi, then across land to Lake Superior, and direct over the lake to one of the Company's **posts on its** northern shore. Thence by a chain of lakes, rivers, and "portages" to York factory, and on northward to Fort Churchill. Of course, at Fort Churchill they had arrived within the range of the great white or polar bear (*ursus maritimus*), who was to be the next object of their "chasse." In the neighborhood of York factory, and even farther to the south, they might have found bears of this species: for the *ursus maritimus* extends his wanderings all round the shores of Hudson's Bay — though not to those of James's Bay further south. The latitude of 55° is his southern limit upon the conti-

nent of America; but this only refers to the shores of
Labrador and those of Hudson's Bay. On the western
coast Behring's Straits appear to form his boundary
southward; and even within these, for some distance
along both the Asiatic and American shores, he is one
of the rarest of wanderers. His favorite range is among
the vast conglomeration of islands and peninsulas that
extend around Hudson's and Baffin's Bays — including
the ice-bound coasts of Greenland and Labrador — while
going westward to Behring's Straits, although the great
quadruped is occasionally met with, he is much more
rare. Somewhat in a similar manner, are the white
bears distributed in the eastern hemisphere. While
found in great plenty in the Frozen Ocean, in its central
and eastern parts, towards the west, on the northern
coasts of Russia and Lapland, they are never seen —
except when by chance they have strayed thither, or
been drifted upon masses of floating ice.

It is unnecessary to remark that this species of bear
lives almost exclusively near the sea, and *by* the sea.
He may be almost said to dwell upon it: since out of
the twelve months in the year, ten of them at least are
passed by him upon the fields of ice. During the short
summer of the Arctic regions, he makes a trip inland —
rarely extending it above fifty miles, and never over a
hundred — guided in his excursions by the courses of
rivers that fall into the sea. His purpose in making
these inland expeditions, is to pick up the fresh-water
fish; which he finds it convenient to catch in the numerous falls or shallows of the streams. He also varies his
fish diet at this season, by making an occasional meal

on such roots and berries as he may find growing along
the banks. At other times of the year, when all inland
water is frozen up, and even the sea to a great distance
from land, he then keeps along the extreme edge of the
frozen surface, and finds his food in the open water of
the sea. Sea-fish of different species, seals, the young
walrus, and even at times the young of the great whale
itself, become his prey — all of which he hunts and cap-
tures with a skill and cunning that appear more the
result of a reasoning process than a mere instinct.

His natatory powers appear to have no limit: at all
events, he has been met with swimming about in open
water full twenty miles from either ice or land. He has
been often seen much farther from shore, drifting upon
masses of ice; but it is doubtful whether he cared much
for the footing thus afforded him. It is quite possible he
can swim as long as it pleases him, or until his strength
may become exhausted by hunger. While going through
the water, it does not appear necessary for him to make
the slightest effort; and he can even spring up above the
surface, and bound forward after the manner of porpoises
or other *cetaceæ*.

If any quadruped has ever reached the pole, it is the
polar bear; and it is quite probable that his range extends to this remarkable point on the earth's surface.
Most certainly it may, if we suppose that there is open
water around the pole — a supposition that, by analogical
reasoning, may be proved to be correct. The daring
Parry found white bears at 82°; and there is no reason
why they should not traverse the intervening zone of
five hundred odd miles, almost as easily as the fowls of

the air or the fish of the sea. No doubt there are polar
bears around the pole; though it may be assumed for
certain that none of them ever attempts to "swarm"
up it, as the white bear is not the best climber of his
kind.

The female of the polar bear is not so much addicted
to a maritime life as her liege-lord. The former, unless
when barren, keeps upon the land; and it is upon the
land that she brings forth her young. When pregnant,
she wanders off to some distance from the shore; and
choosing her bed, she lies down, goes to sleep, and there
remains until spring. She does not, like other hybernating bears, seek out a cave or hollow tree; for in the deso-
late land she inhabits, ofttimes neither one nor the other
could be found. She merely waits for the setting in of a
great snow-storm — which her instinct warns her of —
and then, stretching herself under the lee of a rock — or
other inequality, where the snow will be likely to form a
deep drift — she remains motionless till it has "smoored"
her quite up, often covering her body to the depth of several feet. There she remains throughout the winter,
completely motionless, and apparently in a state of torpor. The heat of her body thawing the snow that comes
immediately in contact with it, together with some warmth
from her limited breathing, in time enlarges the space
around her, so that she reclines inside a sort of icy shell.
It is fortunate that circumstances provide her with this
extra room: since in due course of time she will stand in
need of it for the company she expects.

And in process of time it is called into use. When
the spring sun begins to melt the snow outside, the bear

becomes a mother, and a brace of little white cubs make
their appearance, each about as big as a rabbit.

The mother does not immediately lead them forth from
their snowy chamber; but continues to suckle them there
until they are of the size of Arctic foxes, and ready to
take the road. Then she makes an effort, breaks through
the icy crust that forms the dome of her dwelling, and
commences her journey towards the sea.

There are times when the snow around her has become
so firmly caked, that, with her strength exhausted by the
suckling of her cubs, the bear is unable to break through
it. In a case of this kind, she is compelled to remain in
an involuntary durance — until the **sun** gradually melts
the ice around her and sets her free. Then she issues
from her prolonged imprisonment, only the shadow of
her former self, and scarce able to keep her feet.

The Northern Indians and Eskimos capture hundreds
of these hybernating bears every season — taking both
them and their cubs at the same time. They find the
retreat in various ways : sometimes by their dogs scrap-
ing to get into it, and sometimes by observing the white
hoar that hangs over a little hole which the warmth of the
bear's breath has kept open in the snow.

The hunters, having ascertained the exact position of
the animal's body, either dig from above, and spear the
old she in her bed ; or they make a tunnel in a horizon-
tal direction, and, getting a noose around the head or
one of the paws of the bear, drag her forth in that
way.

To give an account of the many interesting habits peculiar to the polar bear — with others which this species

shares in common with the Bruin family — would require
a volume to itself. These habits are well described by
many writers of veracity, — such as Lyon, Hearne, Richardson, and a long array of other Arctic explorers. It is
therefore unnecessary to dwell on them here — where we
have only space to narrate an adventure which occurred
to our young bear-hunters, while procuring the skin of
this interesting quadruped.

11   P

CHAPTER XLIV.

THE OLD SHE SURROUNDED.

THEY had been for some days on the lookout for a white bear; and had made several excursions from the Fort — going as far as the mouth of the Seal river, which runs into Hudson's Bay a little farther to the north. On all these excursions they had been unsuccessful; for, although they had several times come upon the track of the bears, and had even seen them at a distance, they were unable in a single instance to get within shot. The difficulty arose from the level nature of the ground, and its being quite destitute of trees or other cover, under which they might approach the animals. The country around Fort Churchill is of this character — and indeed along the whole western shore of Hudson's Bay, where the soil is a low alluviom, without either rocks or hills. This formation runs landward for about a hundred miles — constituting a strip of marshy soil, which separates the sea from a parallel limestone formation further inward. Then succeed the primitive rocks, which cover a large interior tract of country, known as the " Barren Grounds."

It is only on the low belt adjoining the coast that the polar bear is found; but the females range quite across to the skirts of the woods which cover the limestone formation. Our hunters therefore knew that either upon the shore itself, or upon the low alluvial tract adjoining it, they would have to search for their game; and to this district they confined their search.

On the fifth day they made a more extended excursion towards the interior. It was now the season of midsummer, when the old males range up the banks of the streams: partly with the design of catching a few freshwater fish, partly to nibble at the sweet berries, but above all to meet the females, who, just then, with their half-grown cubs, come coyly seaward to meet their old friends of the previous year, and introduce their offspring to their fathers, who, up to this hour, have not set eyes on them.

On the present excursion our hunters were more fortunate than before: since they not only witnessed a reunion of this sort, but succeeded in making a capture of the whole family, — father, mother, and cubs.

They had, on this occasion, gone up the Churchill river, and were ascending a branch stream that runs into the latter, some miles above the fort. Their mode of travelling was in a birch-bark canoe: for horses are almost unknown in the territory of the Hudson's Bay Company, excepting in those parts of it that consist of prairie. Throughout most of this region the only means of travelling is by canoes and boats, which are managed by men who follow it as a calling, and who are styled "voyageurs." They are nearly all of Canadian origin — many of them half-breeds, and extremely skilful in the navigation of the lakes and rivers of this untrodden

wilderness. Of course most of them are in the employ
of the Hudson's Bay Company; and when not actually
engaged in "voyaging" do a little hunting and trapping
on their own account.

Two of these voyageurs — kindly furnished by the
chief factor at the fort — propelled the canoe which carried our young hunters; so that with Pouchskin there
were five men in the little craft. This was nothing,
however, as birch-bark canoes are used in the territory
of a much larger kind — some that will even carry
tons of merchandise and a great many men. Along
the bank of the stream into which they had now entered
grew a selvage of willows — here and there forming
leafy thickets that were impenetrable to the eye; but in
other places standing so thinly, that the plains beyond
them could not be seen out of the canoe.

It was a likely enough place for white bears to be
found in — especially at this season, when, as already
stated, the old males go inland to meet the females, as
well as to indulge in a little vegetable diet, after having
confined themselves all the rest of the year to fish and
seal-flesh. The voyageurs said that there were many
bulbous roots growing in those low meadows of which
the bears are very fond; and also *larvæ* of certain insects, found in heaps, like ant-hills — which, by Bruin,
are esteemed a delicacy of the rarest kind.

For this reason our hunters were regarding the land
on both sides of the stream, occasionally standing up in
the canoe to reconnoitre over the tops of the willows, or
peering through them where they grew thinly. While
passing opposite one of the breaks in the willow-grove,

a spectacle came before their eyes that caused them to order the canoe to be stopped, and the voyageurs to rest on their oars.

Alexis, who had been **upon the** lookout, at first did not know what to make of the spectacle : so odd was the grouping of the figures that composed it. He could see a large number of animals of *quadrupedal* form, but of different colors. Some were nearly white, others brown or reddish-brown, and several were quite black. All appeared to have long, shaggy hair, cocked ears, and large, bushy tails. They were not standing at rest, but moving about — now running rapidly from point to point, now leaping up in the air, while some were rushing round in circles ! In all there appeared to be thirty or forty of them ; and they covered a space of ground about as large as a drawing-room floor.

There was a slight haze or mist hanging over the meadow, which hindered Alexis from having a clear view of these animals ; and, through the magnifying influence of this sort of atmosphere, they appeared as large as young oxen. Their form, however, was very different from these ; and from their pointed ears, long muzzles, and full, bunching tails, Alexis could think of nothing else to compare them to but wolves. Their varied colors signified nothing : since in these northern lands there are wolves of many varieties from white to black ; and wolves they really were — only magnified by the mist into gigantic proportions.

Alexis had not viewed them long before perceiving that they were not *all* wolves. In their midst was an animal of a very different kind — much larger than any of them; but what sort of a creature *it* was the young hunter could not make out.

Ivan, who had risen to his feet, was equally puzzled to tell.

It appeared as large as half a dozen of the wolves rolled up into one, and was whiter than the whitest of them; but it looked as if it had a hunch upon its back; and altogether more like a shapeless mass of white bristly hair than a regularly-formed quadruped. It must be an animal, however, as its motions testified; for it was seen to be turning round and round, and at intervals darting forward a pace or two, as if working its way in the direction of the river.

Whatever the animal was, it soon became clear that it was battling with the wolves that surrounded it; and this accounted for the singular movements that these last were making, as well as for their fierce barking and growling that, in confused chorus, filled the air. At intervals, and still louder, could be heard a different sort of cry — shrill and plaintive, like the hinny of a mule — and evidently proceeding not from the wolves, but from the huge white animal which they were assailing.

The voyageurs at once recognized the cry.

" A bear ! — a sea-bear !" exclaimed both together.

One of them stood up, and looked over the plain.

" Yes," said he, confirming his **first** assertion. " An old she it is, surrounded by wolves. Ha! it's her cubs they're after! *Voilà messieurs!* She's got one of them on her back. *Enfant de garce*, how the old beldam keeps them at bay! She's fighting her way to the water!"

Guided by the words of the voyageur, our hunters

now perceived clearly enough that the white object appearing over the backs of the wolves was neither more nor less than a large bear; and that which they had taken for a hunch upon its shoulders was another bear — a young one, stretched out at full length along the back of its mother, and clinging there, with its fore-arms clasped around her neck.

It was evident, also, as the voyageur had said, that the old she was endeavoring to work her way towards the river — in hopes, no doubt, of retreating to the water, where she knew the wolves would not dare to follow her. This was evidently her design : for, while they stood watching, she advanced several yards of ground in the direction of the stream.

Notwithstanding the fierce eagerness with which the wolves kept up the attack, they were observing considerable caution in the conflict. They had good reason : since before their eyes was an example of what they might expect, if they came to very close quarters. Upon the ground over which the fight had been raging, three or four of their number were seen lying apparently dead — while others were limping around, or sneaked off with whining cries, licking the wounds they had received from the long claws of their powerful adversary.

It was rather an odd circumstance for the wolves to have thus attacked a polar bear — an antagonist of which they stand in the utmost dread. The thing, however, was explained by one of the voyageurs; who said that the bear in question was a weak one — half famished, perhaps, and feeble from having suckled her young; and it was the cubs, and not the old bear herself, that the wolves were after — thinking to separate these from their mother, and so destroy and devour them. Perhaps one of them had been eaten up already : since only one could be seen ; and there are always two cubs in a litter.

Our young hunters did not think of staying longer to watch the strange encounter. Their sole idea was to get possession of the bear and her cub ; and with this intent they ordered the voyageurs to paddle close up to the shore and land them. As soon as the canoe touched the bank, both leaped out ; and, followed by Pouchskin, proceeded towards the scene of the conflict, — the voyageurs remaining in the canoe.

CHAPTER XLV.

A WHOLE FAMILY CAPTURED.

THE party had not gone more than a dozen steps from the water's edge, when a new object coming under their eyes caused them to halt. This was another quadruped that at that moment was seen dashing out from the willows, and rushing onward towards the scene of the strife. There was no mistaking the character of the creature. Our hunters saw at a glance that it was a large white bear — much larger than the one surrounded by the wolves. It was, in fact, the male; who, wandering in the thicket of willows — or, more likely, lying there asleep — had not till that moment been aware of what was going on, or that his wife and children were in such deadly danger. Perhaps it was the noise that had awaked him; and he was just in the act of hastening forward to the rescue.

With a shuffling gallop he glided over the plain — as fast as a horse could have gone; and in a few seconds he was close up to the scene of the conflict — to which his presence put an end right on the instant. The wolves, seeing him rush open-mouthed towards them, one and all bolted off; and ran at full speed over the plain, their long tails streaming out behind them. Those that were

11 *

wounded, however, could not get clear so easily; and the
enraged bear, charging upon these, rushed from one to
the other, knocking the breath out of each as he came up
to it, with a single " pat " of his heavy paws.

In less than ten seconds the ground was quite cleared
of the ravenous wolves. Only the dead ones remained
on it; while the others, having got off to a safe distance,
halted in straggling groups; and, with their tails drooping upon the grass, stood gazing back with looks of melancholy disappointment.

Bruin, meanwhile, having settled his affair with the
wounded wolves, ran up to his mate; and throwing his
paws around her neck, appeared to congratulate her upon
her escape! And now did our hunters perceive that
there were two cubs instead of one — that which still
clung fast upon the mother's back, and another which
was seen under her belly, and which she had been
equally protecting against the crowd of assailants that
surrounded her.

Both the little fellows — about as large as foxes they
were — now perceived that they were out of a danger —
which, no doubt, they had perfectly comprehended. That
upon the shoulders of the dam leaped down to the earth;
while the other crawled out "from under;" and both
coming together began tumbling about over the grass,
and rolling over one another in play, the parents watching with interest their uncouth gambols.

Notwithstanding the well-known ferocity of these animals, there was something so tender in the spectacle,
that our hunters hesitated about advancing. Alexis, in
particular, whose disposition was a shade more gentle

than that of his companions, felt certain qualms of compassion, as he looked upon this exhibition of feelings and affections that appeared almost human. Ivan was even touched; and certainly neither he nor his brother would have slain these creatures out of mere wanton sport. They would not have thought of such a thing under ordinary circumstances; and it was only from the necessity they were under of procuring the skin that they thought of it all. Perhaps they would even have passed this group; and taken their chances of finding another, that might make a less powerful appeal to their compassion; but in this they were overruled by Pouchskin. The old grenadier was afflicted by no such tender sentiments; and throwing aside all scruple, before his young masters could interfere to prevent him, he advanced a few paces forward, and discharged his fusil, broadside at the biggest of the bears.

Whether he hit the bear or not, was not then known. Certain it was that he in no way crippled the animal; for, as soon as the smoke had cleared out of his eyes, he saw the huge quadruped part from the side of his mate, and come charging down upon him.

Pouchskin hesitated for a moment whether to withstand the attack, and had drawn his knife to be ready; but the formidable appearance of the antagonist, his immense size, and fierce aspect, admonished Pouchskin that in this case discretion might be the better part of valor, and he yielded to the suggestion. Indeed, the two voyageurs in the canoe were already shouting to all three to run for it — warning them of the danger they were in by the most earnest speech and gesture.

Ivan and Alexis stoood their ground till Pouchskin
had returned to where they were, and then both fired
upon the bear. They may have hit him or not; but the
huge monster showed no sign, and only appeared to
charge forward the faster.

All three together now ran for the boat. It was their
only refuge; for had it been a trial of speed, and much
ground to go over, the bear would certainly have overtaken them; and a few wipes from his paw would have
ended the life of one or the other — perhaps of the
whole trio.

It was fortunate they had the boat to flee to: else
Pouchskin's imprudence, in provoking the bear, might
have led to a fatal termination.

Quick as their legs could carry them they made for
the canoe; and one after the other leaped into it. Without even waiting for them to seat themselves, the two
voyageurs pushed off from the bank, suddenly shooting
the craft out into the middle of the stream.

But this did not stay the pursuit of the infuriated
bear, nor even delay him for a moment.

On reaching the bank, he did not make halt; but,
launching out, sprang down with a plunge upon the water.
Then, stretching his body at full length, he swam direct
after the canoe.

The craft had been turned head down the stream;
and, what with the help of the current and the impulse
of the oars, it swept onward with arrow-like rapidity.
But for all that it soon became apparent that the bear
was gaining upon it — his broad paws enabling him to
swim with the velocity of a fish — while every now and

then he rose above the surface, and bounded forward to a distance of several feet through the air!

The voyageurs plied their paddles with all their skill and energy; there was the dread of death to stimulate them to the utmost exertion of their strength. They knew well, that, if the bear should succeed in coming up with the canoe, he would either mount into it, and drive all of them into the water; or, what was more probable, he would upset the craft, and spill the whole party out of it. In either case, there would be the danger of coming in contact with his claws; and that, they knew, was the danger of death itself.

The hunters were all three busy reloading their guns; and getting ready to fire before the enemy should be up to them.

They were not in time, however. With the motion of the boat, and the constrained attitudes in which it placed them, the loading was a slow process; and, before any of the three had a bullet down, the bear was close astern. Only Ivan had a barrel loaded; and this, unfortunately, was with small shot, which he had been keeping for waterfowl. He fired it, nevertheless, right into the teeth of the pursuer; but, instead of stopping him, it only increased his rage, and roused him to make still greater efforts to overtake the canoe.

Pouchskin, in despair, threw down his gun, and seized upon an axe, that by good luck had been brought in the boat. With this firmly grasped in his hands, and kneeling in the stern, he waited the approach of the infuriated swimmer.

The bear had got close up to the boat — in fact was

within the length of his own body of touching it. Believing himself now near enough, he made one of his prodigious bounds, and launched himself forward. His sharp claws rattled against the birch bark, tearing a large flake from the craft. Had this not given way, his hold would have been complete; and the boat would, in all likelihood, have been dragged, stern foremost, under water. But the failure of his clutch brought the head of the monster once more on a level with the surface; and before he could raise it to make a second spring, the **great** wedge of steel descended upon his crown, and went crashing through his skull.

Almost in the same instant, he was seen to turn over in the water; his limbs moved only with a spasmodic action; he gave a feeble kick or two with his long hind legs; and then his carcass floated along the surface, like a mass of white foam.

It was soon secured, and drawn out upon the bank — for the purpose of being stripped of its snow-white robe.

Our young hunters would have been contented to have left the others alone — neither the female nor her cubs being required by them. But the voyageurs — who were desirous of obtaining the skins of all three on their own account — proposed returning to effect their destruction; **and** in this proposal they were backed by Pouchskin, who had a natural antipathy to all bears.

It ended in the killing of the dam, and the capturing of her cubs alive; for, encumbered as the old she was with her offspring, she was soon overtaken, and fell an easy victim to the volley **of** bullets that were poured into her from all sides at once.

With the skins of the old bears, and the cubs tied in the bottom of the canoe, our hunters started back down stream; but they had scarce parted from the place, before the ravenous wolves returned — not only to devour the carcasses of the bears, but also those of their own comrades that had fallen in the encounter!

CHAPTER XLVI.

THE BARREN GROUNDS.

THE " Barren Ground bear" was next to be sought
for; but to reach the haunts of this animal, a long and
toilsome journey must be made. That tract of the
Hudson's Bay territory known as the " Barren Grounds,"
extends from the shores of the Arctic Sea as far south
as the latitude of the Churchill river; bounded eastward by Hudson's Bay itself, and westward by a chain
of lakes, of which the Great Slave and Athapescow
are the principal.

This immense territory is almost unexplored to the
present hour. Even the Hudson's Bay trappers have
a very imperfect knowledge of it. It has been crossed
in one or two places, and skirted by exploring parties,
but it is still almost a *terra ignota*, except to the four
or five tribes of Indians who dwell around its borders,
and the Esquimaux, who venture a little way into it
along the coast of the Arctic Sea.

Before proceeding to hunt the Barren Ground bear,
let us say a word about his species. By writers, both
old and modern, he has been variously classed. Even
the ablest naturalist who has written about him is puzzled as to his species. We speak of Sir John Rich-

ardson, the companion of the lamented Franklin, and
himself one of the great men of the earth. Sir John
first regarded this bear, though very doubtfully, as a
variety of the *ursus americanus*, or American black
bear. Later observations influenced him to change this
opinion; and again with modest doubtfulness — characteristic of the man — he suggests his being a variety
of the *ursus arctos*.

We shall make bold to affirm that he is a variety
of neither; but a distinct species of bear.

We shall give our reasons — and first, as to his dis-
tinctness from the *ursus americanus*. He is not like
the latter, either in color, shape of body, bulk, profile,
physiognomy, length of feet or tail. In all these
respects he bears a greater resemblance to the *ursus
arctos*, or even to his nearer neighbor, the grisly (*ursus
ferox*). He differs from both these, however, in other
points — as will presently be seen. Again, he is of a
fiercer disposition than the black bear, and more dangerous to the hunter — almost as much so as the grisly,
and quite as much as the brown. Moreover, he dwells
in a country in which the black bear could not make
his home. To the existence of the latter, the forest is
essential; and he is never found far out of it. It is
not the higher latitude that keeps him out of the Barren
Grounds, but the absence of timber. This is proved
by the fact of his being found quite as far northward
as any part of the Barren Grounds, but where the
limestone formation favors the growth of trees; whereas, among the primitive rocks to the north of Nelson
river, the black bear does not exist — the very region

Q

that appears most favorable to the existence of the
Barren Ground species — who cares not for trees, and
cannot climb them.

Still another material difference may be pointed out.
The black bear, in his normal state, is altogether frugivorous — a true vegetable feeder. The other is carnivorous and piscivorous — at one season killing and eating marmots and mice, at another frequenting the sea-
coast and subsisting upon fish. In a word, the two
bears are as unlike as may be — they are distinct
species.

To compare the Barren Ground bear with the *ursus
arctos.* The former is certainly much more like this
species, than he is to the *ursus americanus ;* but again
we encounter notable points of difference; and were it
not for a certain resemblance in color, it is possible the
two kinds would never have been brought into comparison. It is easy, however to prove them also distinct
species — by simply observing that their habits are altogether unlike. The *ursus arctos* is a *tree-climbing
wood-bear:* the Barren Ground species is not. The
former prefers a vegetable diet — the latter likes better fish, flesh, and insects — though he will also fill his
stomach with a farrago of vegetable matters.

But to say nothing of the very different habits of the
two animals, there is a yellowish tinge over the fur of the
American species that is not observed in the brown bears
of European countries — except, perhaps, in those of the
Pyrenees — and at certain seasons this tinge turns so
pale, as to give a whitish appearance to the animal:
hence, by the Indians, they are often termed " white
bears."

It is, besides, altogether improbable, that the brown bear of Europe should turn up in the " Barren Grounds" of the Hudson's Bay territory — an isolated, treeless tract — quite unlike his habitat in the Old World ; and to which no line of migration could be traced with much probability. We might suppose such a migration through Siberia and Russian America ; and certainly there is some probability in this view : for although it has been hitherto stated that the Barren Ground bear is only found within the limits of the peculiar district so called, it is very certain that his range extends beyond these boundaries. The brown bear of Russian America and the Aleutian Islands appears to be identical with this species ; and there is a suspicion that the brown species of Kamschatka is no other than the Barren Ground bear of the Hudson's Bay. The fishing habits of the former go some ways towards an identification of the two species — at the same time separating both from the *ursus arctos* of Scandinavia.

It needs hardly to be argued, that the Barren Ground bear is quite a distinct animal from the grisly — though writers have often confounded them. They are different in size and color. Though the grisly is sometimes brown, it is always with a mixture of white-tipped hairs ; but the most essential distinction is to be found in the greater ferocity of the latter, and his far longer and more curving claws. Many other points might be mentioned — showing them to be animals of two separate species — besides, their range is altogether distinct.

The Barren Ground bear, then, is not the *ursus arctos, americanus,* or *ferox.* What then? Has he received

no specific name from the naturalists? Not yet. Alexis,
however, bestowed one upon him. He named him after
the man who has given the clearest account of his country
and his habits; and whom Alexis deemed most worthy
of the honor. In his journal we find the record. There
it is written, that the Barren Ground bear is the *ursus
Richardsonii*.

CHAPTER XLVII.

BRUIN TAKING A BATH.

To seek the haunts of this new species of bear, I have said that our hunters would have a long journey to make, even so far as the Great Slave Lake — for although the Barren Grounds extend many degrees to the south of this water, the *ursus Richardsonii* rarely wanders to a lower latitude. Upon the shores of the Slave Lake, however, they would be certain to encounter him, and thither they repaired.

They were fortunate in the time of the year. The annual "brigade" of boats belonging to the Great Fur Company was just setting out from York Factory, for Norway House on Lake Winnipeg; and thence a division of it would proceed to the posts still further northward — on Lake Athapescow and the waters of the Mackenzie River — passing through the Slave Lake itself. Their object, of course, in their annual journey is to distribute at the fur-stations, the goods, brought from England by the Company's ships, and in return bring back the peltries collected throughout the winter.

With the brigade, then, went our hunters; and after enduring, in common with the others, the hardships and perils incidental to such a long inland voyage, they at length found themselves at the point of their destination, Fort Resolution, on the Great Slave Lake, near the mouth of the river bearing the same appellation. The canoe of an Indian fisherman — of which there are many dwelling around the shores of this great inland **sea** — was soon pressed into service; and with the fisherman (**who** of course was a hunter also) for their guide and companion, they could **make** convenient excursions along the shores of the lake, land whenever they pleased, and search for Bruin in the localities where he was most likely to be encountered. In this they were assisted by their hired guide; who was not long in putting them upon the trail **of a bear**. In fact, in the very first excursion which they made, one of the true breed was discovered and captured.

The circumstances attending his capture were of no very particular interest; but as they illustrate one of the habits of this species, we shall give them as recorded in the journal of Alexis.

They were paddling gently along the shore — through water that was as calm as a pond — when, at a great distance ahead of them, the Indian observed a slight rippling upon the surface, and pointed it out. It was not caused by the wind; for there was not a breath stirring at the time; and it was not like the whitish curl which a breeze casts upon the surface of water. It resembled more a series of little wavelets, such as proceed from a stone plunged into a deep pool, or from a disturbance of the water caused by the movements of some animal. The Indian said that it was a bear: though there was no bear, nor any living thing in sight!

As the canoe moved nearer, our hunters perceived that there was an indentation on the shore — a little creek or bay out of which the ripples were proceeding. The guide knew that there was such a bay; and believed that the bear would be found somewhere within it, swimming about in the water.

The hunters did not stay to inquire the reason why Bruin should be thus bathing himself. There was no time: for just at that instant the Indian beached his canoe; and desired them all to debark and follow such further instructions as he might give them. Without hesitation they accepted his invitation; resolved to act according to his counsel.

The Indian, after making his boat fast, took the route inland, followed by the other three. After going some three or four hundred yards, he turned to the left, and conducted the party around the shore of the bay — which trended in a semicircular or horseshoe shape. He did not take all of them around; but only one, whom he stationed on the opposite side. This was Pouchskin. Ivan he had already placed on the nearer side, and Alexis at the bottom — so that they were thus set at the three angles of a triangle, nearly equilateral.

On assigning to each of them his station, the Indian further instructed them to creep forward among the bushes — which still separated them from the water — and to do so without making any noise, till they should hear a "whoop" from himself. This would be the signal for them to show themselves around the edge of the bay — in the water of which the Indian hunter was confident a bear was bathing himself. He himself returned to his canoe.

Agreeably to his instructions, the three hunters crawled forward — each on his own line of approach, and all observing the greatest caution and silence. As soon as their eyes rested upon the water, they perceived the correctness of the Indian's conjecture. A bear there was, sure enough!

They saw only his head; but this was sufficient for Bruin's identification: since no similar cranium could have been encountered in such a place.

As the Indian had apprised them, the bear was swimming about in the bay; but for what purpose it was at first difficult to make out. To their astonishment, he swam with his mouth wide open — so that they could see the interior of his great encarmined palate, while his long tongue flapped out at intervals, and appeared to sweep the surface of the water. At intervals, too, he was seen to close his mouth — the huge jaws coming together with a "clap-clap," the noise of which could be heard echoing far over the lake!

He did not go long in one course; but ever and anon kept turning himself, and quartering the bay in every direction.

It was a long time before the spectators could find any explanation of these odd manœuvres on the part of the bear. They might have fancied he was merely taking a cool bath to refresh himself: for the day was exceedingly hot, and the air was filled with mosquitoes — as our hunters had already learnt to their great discomfort. It might have been to get rid of these tormentors that Bruin had submerged his body in the water; and so Pouchskin concluded, and also Ivan — though both were puzzled by the odd behavior of the bear, in swimming open-mouthed, and at intervals snapping his jaws as he did. Alexis, however, was a better reasoner; and soon discovered the why and the wherefore of these mysterious demonstrations. Alexis saw that the surface of the water was thickly coated with something; and, on scrutinizing it more closely, he made out this something to be a swarm of insects. There appeared to be more than one species of them — two indeed there were — both about the size of ordinary gadflies; but altogether different from each other in color and habits. One was a sort of water-beetle that swam near the surface; while the other was a winged insect that occasionally rose into the air, but more generally crawled along the water — making short runs from place to place, then stopping a moment, and then darting on again. The whole surface of the bay — and even out for some distance into the lake — fairly swarmed with these creatures; and it was in pursuit of them that Bruin was whisking his tongue so rapidly about, and bringing his jaws together in such sonorous concussion. The animal was simply indulging in a favorite meal — which in summer is furnished him not only on the shores of the Great Slave Lake, but most of the smaller lakes throughout the Barren Grounds.

Alexis had scarce finished making the observation, when a loud "whoop" was heard from the direction of the lake; and almost at the same instant the canoe of the Indian was seen shooting through the water, right for the entrance of the bay!

Obedient to the signal, the three hunters rushed out from their cover, and ran forward upon the beach —

12

each holding his gun in readiness to fire. The bear, seeing himself thus suddenly and unexpectedly surrounded,
at once gave over his fly-trapping; but, irresolute in
which direction to retreat, he turned round and round in
the water, first swimming a bit one way and then another.
At length, rearing himself high above the surface, and
showing his sharp teeth, he uttered a deep growl of rage,
and dashed recklessly towards the shore.

It was to Ivan's side he first directed himself; but Ivan
was upon the watch; and, advancing close to the edge of
the water, he took aim and fired.

His bullet struck **the** bear right upon the snout, and it
appeared to have spun him round — so quickly was he
seen heading in the opposite direction.

It was now Pouchskin's turn; and in a second after
the loud report of the grenadier's gun went booming over
the lake, while the ball splashed the water right into the
eyes of the bear. Though it did not hit any part of his
body, it had the effect of half turning him — so that he
now swam towards Alexis, stationed at the bottom of
the bay.

Alexis took the matter more coolly. There was a
convenient tree behind — to which he intended to retreat
in case of missing — and this influenced him to hold his
ground, till the bear should come near enough to insure
a certain aim.

The bear swam straight on, until within some ten
yards of where Alexis was standing; when all at once
he appeared to take the rue, and was turning off to one
side. This was just what Alexis desired: it brought
the head of the animal broadside towards him, and,

taking steady aim, he planted his bullet a little under the left ear.

It was a dead shot. The huge creature, loaded with fat, sank instantly to the bottom; but fortunately the water was shallow; and the Indian, now coming in with his canoe, soon fished up the carcass, and towed it out upon the beach — where its fur coat was stripped off in a trice.

CHAPTER XLVIII.

THE GREAT GRISLY.

THE grisly bear (*ursus ferox*), the fiercest and most formidable of the ursine family, was the next to be captured and skinned.

The range of the grisly, though wider than that of the Barren Ground bear, is still not so extensive as that of the *ursus americanus*. The great chain or cordillera of the Rocky Mountains may be taken as the *axis* of his range — since he is found throughout its whole extent, from Mexico to its declension near the shores of the Arctic Sea. Some writers have asserted that he is confined to these mountains, but that is an error. To the west of them he is encountered throughout all the countries lying between the Rocky Mountains and the Pacific coast — wherever circumstances are favorable to his existence ; and to the east he extends his wanderings for a considerable distance into the great plains — though nowhere so far as to the wooded countries near the meridian of the Mississippi. In these the black bear is the only forest-ranger of the family.

Woods are not the favorite haunt of the grisly bear ; and although in youth he can make a sort of scramble up a tree, when full grown his enormous claws — always blunted at the tips — hinder him from climbing. Low, bushy thickets, with open glades intervening — and especially where the underwood consists of berry-bearing bushes — are his chosen retreats. He often sallies out into the open ground ; and on those prairies where grows the *pomme blanche*, or "Indian turnip" (*psorolea esculenta*), he may be seen tearing up the earth with his claws, and leaving it turned into furrows — as if a drove of hogs had been "rooting" the ground. On the bottoms of the streams he also digs up the "kamas" root (*camassia esculenta*), the "yampah" (*anethum graveolens*), the "kooyah" (*valeriana edulis*), and the root of a species of thistle (*circium virginianum*). Many species of fruits and berries furnish him with an occasional meal ; and the sweet pods of the mesquites (species of *acacia*), and the cones of the piñon-tree (*pinus edulis*) form portions of his varied larder.

He does not, however, confine himself to a vegetable diet. Like most of his kind, he is also carnivorous, and will dine off the carcass of a horse or buffalo. The latter animal, notwithstanding its enormous bulk and strength, frequently falls a prey to the grisly bear. The long masses of hair that hang over the eyes of the buffalo hinder it from perceiving the presence of an enemy ; and, unless warned by the scent, it is easily approached. The bear, knowing this, steals up against the wind ; and, when within safe distance, springs upon the hind quarters of the ruminant, and cramping it in his great claws, succeeds in dragging it to the ground. He is even able to transport the huge carcass to a considerable distance — for the purpose of concealing it in some thicket, and devouring it at his leisure.

The grisly bear is more like to the brown bear of
Europe than to any other species of the genus. His fur
is long and shaggy — not presenting the even surface
which characterizes the coat of the black bear. It is
generally of a dark-brown color — the hair being whitish
at the tips, more especially during the summer season,
when it becomes lighter colored. The head is always
of a grizzled gray; and it is this appearance that has
obtained for the animal its specific name. There are
brown, reddish-brown, bay or cinnamon-colored, and
white-breasted varieties of the black bear; but the Indians can distinguish all these from the true grisly at
a glance. In all of the latter, where there are white
hairs intermingled with the fur, it is always observable
that these odd hairs are white to the roots; whereas the
hoary appearance of the grisly is caused by only the tips
of the hair being white. This characteristic is constant;
and would of itself justify a distinction being made between the species; but there are many other points of
greater importance. The ears of the grisly are shorter,
more conical, and set wider apart than in either the
ursus americanus or *arctos.* His claws áre white, arched,
far longer, and broader than those of the other bears —
their greatest breadth being across their upper surface.
Underneath they are chamfered away to a sharp edge;
and projecting far beyond the hair of the foot, they cut
like chisels when the animal strikes a blow with them.
His huge paw is both broader and longer than that of
other bears; while his tail, on the other hand, is short
and inconspicuous — being completely buried under the
fur of his buttocks. So characteristic is this appendage

for its extreme shortness, that it is a standing joke among the Indians — when they have killed a grisly bear — to desire any one unacquainted with the animal, to take hold of its tail!

This appendage in the *ursus americanus* and *ursus arctos* is conspicuous enough ; and in the Barren Ground bear is still longer than in either.

There could be no possibility of mistaking an old or full-grown grisly for any of the kindred species. Both in size and aspect he is different. It is only in the case of young or half-grown specimens where a mistake of this kind is likely to be made. The enormous size of the old males — often weighing 1,000 lbs., and quite equalling the largest individuals of the *ursus maritimus* — renders them easy of identification ; though it is certain that under favorable circumstances the *ursus arctos* often attains to a similar bulk.

In ferocity of disposition, however, in carnivorous inclination, and in strength and power to carry out his mischievous propensities, no bear, not even the *ursus maritimus*, appears to be a match for this monster of the Rocky Mountains. The hunter never thinks of attacking him, unless when assisted by a number of his comrades ; and even then it may be a fatal encounter for one or more of them. Were it not for the advantage obtained by their being mounted on horseback, the grisly would always have a wide berth given him ; but fortunately this fierce quadruped is unable to overtake the mounted hunter — although he can easily come up with a man on foot.

As to fearing or running away from a human antago-

nist, the younger grislies may sometimes do so; but when
an old male has been attacked the case is quite different.
A full-grown individual will stand his ground against a
crowd of assailants — charging from one to the other,
and showing fight so long as there is breath in **his** body.

The number of Indian and white hunters who have
either been killed or badly mutilated by grisly bears
is almost incredible. Were it not that these men are
usually mounted on good horses the list would have been
still greater; and his intended victims often find another
means of escaping from his claws — by taking to a tree.

Fortunate it is that nature has not bestowed upon the
grisly the power of tree-climbing; else many a pursued
hunter, who has succeeded in gaining the branches of a
friendly cottonwood, might have found his refuge anything but a secure one.

In fact, climbing into a tree — when one can be
reached — is the common resource of all persons pursued by the grisly bear; and by this means did our
hunters themselves escape from a brace of infuriated
grislies, while engaged in hunting these formidable
animals.

CHAPTER XLIX.

A FUR-TRADER'S FORT.

HAVING settled their accounts with Bruin of the Barren Grounds, our travellers proceeded down the Mackenzie river to the Hudson's Bay post of Fort Simpson. Thence they ascended a large tributary of the Mackenzie, known as the "River of the Mountains," — or as the Canadian voyagers call it, *Rivière aux Liards*. This large stream has its sources far beyond the highest peaks of the Rocky Mountains : thus exhibiting the curious phenomenon of a river, breaking through a chain of mountains in a transverse direction ; though the same occurs in several other parts of the Rocky Mountain range, and also in the Andes of South America. On the *Rivière aux Liards* the Hudson's Bay Company have several posts — as Forts Simpson, Liard, and Halkett — the last mentioned being far up among the mountains. Westward again, upon the Pacific side, they have other trading-stations — the most important of which is that of Pelly's Banks, situated at the junction of Lewis and Pelly rivers. These rivers, after joining, run into the Pacific, not far from Mount St. Elios — long noted as a landmark to the navigators of the North Pacific ocean.

12 *   R

From Fort Halkett, a route has been established to the post at Pelly's Banks by means of Dease's river — which is one of the effluents of the *Rivière aux Liards* — and, partly by canoe navigation and partly by "portage," the continent can be crossed in this northern latitude. From Pelly's Banks to the Pacific coast the route is still easier — for not only do the Russians visit these parts, but there are native Indian traders who go twice every year from Pelly's Banks to Sitka — the entrepôt of the Russian Fur Company — and the Lynn channel, a little to the north of Sitka, is also visited by the steamers of the Hudson's Bay Company itself.

Our travellers would therefore have no difficulty in reaching Sitka; and thence crossing to the peninsula of Kamschatka, on the Asiatic coast. On their way over the Rocky Mountains, they would be certain to fall in with the grisly; and in the countries lying along the Pacific, they could obtain that variety of the *ursus americanus,* known as the "cinnamon bear" — for it is to the west of the Rocky Mountains — in California, Oregon, British Columbia, and Russian America — that this spice-colored species is most frequently met with.

A party of fur-traders and trappers were just starting from Fort Simpson to carry supplies up to the posts of Liard and Halkett; and along with them our travellers went.

On reaching the last-named station, they came to a halt, for the purpose of hunting the grisly.

They were not long in starting their game — for this fierce monster of the mountains is far from being a scarce animal. In fact, in those districts which they choose for their "beat," the grisly bears are more numerous than most other quadrupeds; and not unfrequently half a dozen or more of them may be seen together. It is not that they are *gregarious;* but simply, that, being in considerable numbers in a particular neighborhood, accident thus brings them together. To see troops of four associating together is very common; but these are merely the members of one family — male, female, and yearling cubs — for two is the number of the progeny — the grisly bear in this respect resembling his congener of the *ursus maritimus,* and differing as essentially from the black and brown bears — with whom three is the usual number of cubs at a birth.

There are good reasons why the grisly bears are not in much danger of being exterminated. In the first place, their flesh is of inferior quality. Even the Indians will not eat it; while they relish that of the black species. Secondly, their robe is of scarce any value, and fetches but a trifling price in the fur market. Thirdly — and perhaps the most powerful reason of all — is that the hunter cares not to risk his life in an encounter with these animals, knowing that there is no adequate reward for such risk. For this reason "Old Ephraim" — as the trappers jocosely style the grisly — is usually permitted to go his way without molestation; and, therefore, instead of being thinned off by an exterminating chase — such as is pursued against the buffalo, or even the black bear, whose robe is marketable — the grisly maintains his numerical strength in most places where he is found.

At Fort Halkett — in consequence of a scarcity of hands, and the great pressure of business, in forwarding

the brigade onward to the Pelly Station — our young
hunters were unable to obtain a guide; and therefore
started out for the chase alone — Pouchskin, of course,
being one of the party.

The trading-post of Fort Halkett being situated in the
midst of the wildest region — without any cultivated
ground or other settlement around it — they would not
have far to go before finding a grisly. Indeed, they were
as likely to meet with one in sight of the Fort as anywhere else; and from the moment of passing through
the gate of the stockade they were on the lookout.

They had not the good fortune, however, to meet with
one so very easily, for although they came upon the traces
of bears, and saw numerous signs of them, they could
not set eyes upon them; and returned from their first
excursion rather disheartened with their day's work.

In one thing, however, they had their reward. They
had succeeded in shooting one of the rarest animals of
America, a creature only met with in the more northern
districts of the Rocky Mountains — that is, the "Rocky-
Mountain goat" (*capra americana*). This rare quadruped — whose long, snow-white, silky hair renders it
one of the most attractive of animals — is a true wild
goat; and the only species of the genus indigenous to
America. It is about the size of the common domestic
breeds, and horned as they; but the shining hair over its
flanks and body is frequently so long as to hang down
almost to its hoofs, giving the animal the appearance of
having a much heavier body and much shorter legs than
it really has. Like the ibex **of** Europe, it is only met
with on the loftiest summits of the mountains, upon

peaks and cliffs inaccessible to almost every other quad-ruped — the mountain sheep alone excepted. It is much shyer than the latter, and far more difficult of approach — the consequence being, that its beautiful skin, though highly prized, and commanding a good price, is but rarely obtained, even by the most expert hunters.

Having succeeded in bringing down one of these precious animals, our young hunters were satisfied with their day's work — almost as well as if it had been a grisly they had killed.

On their second day's excursion, however, this feat was also accomplished — as we shall now proceed to relate.

CHAPTER L.

TREED BY OLD EPHRAIM.

THEY had got about a mile from the Fort; and were proceeding cautiously along through a hilly country, where thicket-like groves grew, interspersed with patches of open ground, forming park-like scenery. There are many scenes of this character in the valleys of the Rocky Mountains; and in the more northern latitudes these groves often consist of berry-bearing bushes — such as wild currants, bird and choke-cherries, the *amelanchier* and *hippophäe canadensis*. Of all these fruits the grisly bear is known to be exceedingly fond; and as the thickets among which our hunters had entered contained many trees of the above kinds — at that season drooping under their ripe fruit — it was but reasonable to expect they might find some of the grislies engaged in gathering them. They had been told at the fort that this was a favorite browsing-place of the bear; and, as they passed along they had evidence of the correctness of the information by seeing the cherry-trees with their branches broken — and some of the stems pulled down into a slanting position, — evidently done by the bears to enable them to get conveniently at the fruit. From the trees that had been treated in this rough manner all the fruit had been stripped off as clean as if a party of "cherry-pickers" had passed that way.

The ravages exhibited a very recent sign. Most of them must have been done within a week; and one tree looked as freshly torn, as if it had been pulled about that very morning.

Of course, with such indications before their eyes, our hunters were advancing on the *qui vive* — not knowing the instant that Bruin might break out.

It would not be correct to say that they were proceeding with caution. Had they been sufficiently cautious, they would not have been there *afoot*. Of course they were on foot — since no horses could be procured in these parts. To go afoot in pursuit of such game as grisly bears was the height of indiscretion; and the traders had told them so; but they made light of what they had been told, for two reasons, — first, because it was absolutely necessary they should kill a grisly and strip him of his skin; and secondly, because our young hunters, Pouchskin as well, had but a very indefinite idea of the risk they were running. They had heard that the grisly was one of the fiercest of its kind; but because it was called a bear, and they had now hunted and killed so many other bears, they fancied this one might be as easily conquered as any of its congeners. They had heard that these animals often turn tail and run away at sight of man; but these stories are deceptive. The bears that do so are either juvenile grislies or brown individuals of the *ursus americanus* — which are often mistaken for the grisly.

With "Old Ephraim" himself the case is quite differ-

ent, as we have already said. On sight of a human
enemy, instead of running away, the grisly more frequently runs towards him, charging forward with open
mouth, and often without having received the slightest
provocation.

Of this fact our hunters had **proof almost upon the** instant. They had entered a wide tract, sparsely covered
with trees; but such small trees, and so thinly standing
over the ground, that the hunters might have fancied
them to have been planted; and that they were entering
within the boundaries of some old orchard. The tract
thus characterized **was** about five or six acres in superficial extent; **and** surrounded **by** the same kind of cop-
pice that covered most of the face of the country.

Under the thin trees there was neither underwood, nor
long grass; and they could see between their trunks **in**
every direction, **to** the edge of the jungle that grew
around.

While walking quietly along, a singular noise reached
their ears, that caused them suddenly to halt in their
tracks. It caused them to turn also: for the noise appeared to come from behind them. It resembled the
hurried breathing of a person badly afflicted with asthma;
but so much louder, that if it had proceeded from human
lungs, they could only have been those of an asthmatic
giant!

It was, in reality, a gigantic creature that produced the
noise: since it was neither more nor less than a grisly
bear. Not one alone, but a brace of these monstrous
animals — a male and female, no doubt — were seen at
that moment by the edge of the thicket, out of which the

hunters had just emerged. Both were standing on their hind limbs, and both uttering the strange snuffing noise that had attracted attention to them. Other noises were now mingled with these — sharp querulous grunts — and, by the gestures which the bears were making, it was evident they not only saw the three hunters in the open ground, but were reconnoitring them, perhaps with an intention to make an attack upon them!

Our hunters were quite taken aback. They had expected, at least, to have been allowed the initiative in any conflict that might occur; but they now saw that, instead of being the assailing party, they were likely to be the assailed!

They had no time for deliberation; for the brace of bears, apparently having satisfied themselves with their threatening demonstrations, dropped down on all-fours, and came galloping onward — almost as fast as horses could have done!

The three hunters fired at once; and not without effect: for one of the bears fell to their shots. It was the smaller one, and that which had been foremost. Acting without concert, they had all aimed at the same animal — choosing that which was nearest; and this was unfortunate, for had some one of them sighted the other and bigger bear, they might have given him a wound that would have, at least, crippled him.

As it was, he had neither been shot at, nor touched; and the fall of his mate — for it was the male who survived — now so completely exasperated him, that he rushed on with the full determination to deal death among the enemies who had bereaved him.

It was fortunate that he stopped a moment over his
fallen companion. He did so as if to convince himself
that she was dead. It was only for an instant; but a
precious instant that was to all three of the hunters. It
gave them sufficient time to take to a tree — each springing up to the one that was most convenient. Alexis and
Ivan being young and nimble, easily accomplished this
feat; but it cost Pouchskin an effort; and he came very
near making it in vain. He had got his arms over a
branch, and was drawing his great booted legs after him;
but, before he could raise them to a sufficient height, the
bear had arrived upon the ground, and reared upward to
seize him.

Ivan and Alexis uttered a simultaneous shout of alarm.
They saw the shaggy fore-arms of the quadruped doubled
around the legs of their faithful follower; and were looking to see Pouchskin in another moment pulled down
from the tree. What was their delight, as well as astonishment, on seeing the bear fall "slap" back to the earth
— with one of the ex-grenadier's great boots fast clutched
between his paws — while Pouchskin himself was seen
gliding upward to the top branches of the tree!

A shout of joy followed the cry of alarm to which
they had just given utterance; and without another word
all three hastened to reload their guns.

Meanwhile the disappointed bear appeared determined
to revenge himself on the boot; and for some seconds
continued to tear it — both with teeth and claws — till
nothing of its original shape remained. Then, scattering
the fragments over the ground, he desisted from this idle
employment; and rushed back to the trunk of the tree

up which Pouchskin had climbed. He knew — from having often made the experiment — that he could not climb it; nor did he attempt to do so; but seizing the slender trunk in his powerful grasp, he shook the tree backward and forward, as if endeavoring to drag it up by the roots or throw it to the ground.

For some time our hunters were not without apprehensions that he might succeed. The tree was not bigger than an ordinary pear-tree; and its trunk vibrated from side to side, and bent over to such an extent, that its roots could be heard cracking beneath the ground.

Pouchskin, far up in the top, was tossed backward and forward — as if he had been a shuttlecock between two battledoors — and it was just as much as he could do to keep his hold among the branches, much less finish the loading of his fusil, which he had only half accomplished when the rocking began. Had he been alone, his position would have been one of great danger: for no doubt, in process of time, the bear would have torn down the tree. But the efforts of Bruin were brought to a sudden termination: for Ivan and Alexis, having now reloaded, took careful aim, and sent both their bullets into the body of the beast. One of the shots must have hit him in a mortal part: since, on receiving it, the bear let go his hold, dropped down from his erect attitude, and doubling himself up at the bottom of the tree, looked as if he had suddenly gone to sleep! But the red stream, pouring out from his still distended jaws, told that it was the sleep of death that had overtaken him.

Our hunters, assured that both bears were dead, now descended from their respective perches; but the sight of Pouchskin, with one leg in stocking, and the other buried up to the thigh in a great horse-skin boot, would have been too much for the gravity of a judge, and his young masters were once more merry at his expense.

Having skinned the bears, they returned to the fort with their spoils — to the no slight astonishment of some of the old trappers stationed there. They could scarce believe that these young strangers were capable of accomplishing such a feat as the conquest of a couple of full-grown grislies. The thing had been done, however — as the trophies testified — and it is needless to say that our hunters, by this gallant action, gained golden opinions from the "mountain men."

They had no desire, however, to try another contest of the kind. They had become perfectly satisfied of the great peril to be expected in an encounter with " Old Ephraim;" and were only too well pleased of having it in their power, on all future occasions, to imitate the example of other travellers, and give the grisly a "wide berth."

Indeed, they would **have had no** opportunity, had they desired it, to hunt the **bear** any longer in that neighborhood: for the "boat" brigade, with which they were travelling, started the next day for Fort Pelly; and it was necessary for them to accompany it, as the journey could not otherwise be accomplished.

They arrived at this last-named place in safety; and, with some native traders, that chanced to be at the fort, they were enabled **to** proceed onward to the Russian settlement of Sitka — where the magic cipher which Alexis carried in his pocket procured them the most hospitable treatment that such a wild, out-of-the-way place could afford.

They had been fortunate, upon their route, to procure a skin of the "cinnamon" bear — as well as one of black color with a white breast, both of which Alexis was able to identify as mere varieties of the *ursus americanus*. These varieties are sometimes seen to the east of the Rocky Mountains; but they are far more common throughout the countries along the Pacific — and especially in Russian America, where the cinnamon-colored kind is usually termed the "red bear." They occur, moreover, in the Aleutian islands; and very probably in Japan and Kamschatka — in which country bears are exceedingly numerous — evidently of several species, confusedly described and ill identified. Unfortunately, the Russian naturalists — whose special duty it has been to make known the natural history of the countries lying around the North Pacific — have done their work in a slovenly and childlike manner.

CHAPTER LI.

THE KAMSCHATDALES.

THE bear of Kamschatka had to be skinned next. But it was necessary to catch one before he could be skinned; and also necessary to go **to** Kamschatka before he could be caught. To get to Kamschatka was not so difficult as it may sound to the ear. Our travellers were just in the place from which it was possible to **proceed** direct to this Asiatic peninsula. Vessels belonging to the Russian Fur Company every year collect the furs along the northwest coast of America, and among the Fox and Aleutian islands — Sitka being their port of rendezvous. **Thence proceeding to the harbor of St.** Peter and **St. Paul** (Petropaulouski), on the coast of Kamschatka, they complete their cargoes with the "skin crop" **that** during the **winter** has been collected throughout the peninsula. Thence to China a portion of these furs are taken — especially skins of the sable, which the Chinese mandarins use extensively **for** trimming their costly robes; and for which teas, silk, lacquer-ware, and other articles **of** Chinese manufacture are given in exchange.

The Japanese also, and other wealthy Oriental nations, buy up quantities of costly furs; but by far the greater portion of this produce is consumed by the Russians themselves — in whose cold climate some sort of a fur coat is almost a necessity. Even most of the furs collected by the Hudson's Bay Company find their way into Russia: for the consumption of these goods in Great Britain is extremely limited, compared with that of many other articles *de luxe*.

In the fur-ship our travellers proceeded from Sitka to the port of Petropaulouski, which is situated on Avatcha bay, near the southern end of the peninsula.

As Avatcha bay is nearly land-locked, it forms one of the most sheltered harbors on that side of the Pacific; but unfortunately during winter the bay freezes over; and then ships can neither get into nor out of it.

The vessel which carried our adventurers arrived at Petropaulouski late in the spring; but, as the winter had been unusually prolonged, the bay was still blocked up with ice, and the ship could not get up to the little town. This did not hinder them from landing. Dog-sledges were brought out upon the ice by the inhabitants; and upon these our travellers were carried to the town, or "ostrog" as it is called — such being the name given to the villages of Kamschatka.

In Petropaulouski, many curious objects and customs came under the observation of our travellers. They saw no less than three kinds of houses — first, the "isbas," built of logs, and not unlike the log-cabins of America. These are the best sort of dwellings; and belong to the Russian merchants and officials, who reside there — as well as to the Cossack soldiers, who are kept by the Russian Government in Kamschatka.

The native Kamschatdales have two kinds of houses
of indigenous architecture — one for summer, the " balagan," and another to which they retire during the winter,
called the " jourt." The balagan is constructed of poles
and thatch upon a raised platform — to which the Kamschatdale climbs up by means of a notched trunk of a
tree. There is only one story of the house itself — which
is merely the sloping thatched roof — with a hole in the
top to give passage to the smoke — and resembles a
rough tent or hayrick set upon an elevated stand. The
space under the platform is left open; and serves as a
storehouse for the dried fish, that forms the staple food
of all sorts of people in Kamschatka. Here, too, the
sledges and sledge-harness are kept; and the dogs, of
which every family owns a large pack, use this lower
story as a sleeping-place.

The winter house or " jourt," is constructed very differently. It is a great hole sunk in the ground to the
depth of eight or ten feet, lined round the sides with
pieces of timber, and roofed over above the surface of
the ground — so as to look like the rounded dome of a
large bake-oven. A hole at the apex is intended for the
chimney, but it is also the door : since there is no other
mode of entrance into the jourt, and the interior is reached
by descending a notched tree-trunk — similar to that used
in climbing up to the balagan.

The curious fur dresses of the Kamschatdales ; their
thin, yellowish-white dogs, resembling the Pomeranian
breed ; their dog-sledges, which they use for travelling in
winter ; the customs and habits of these singular people ;
all formed an interesting study to our travellers, and en-

riched their journal with notes and observations. We find it recorded there, how those people spend their time and obtain their subsistence. Very little agriculture is practised by them — the climate being unfavorable to the growth of the cereals. In some parts barley and rye are cultivated; but only to a very limited extent. Cattle are scarce — a few only being kept by the Russian and Cossack settlers; and horses are equally rare, such as there are belonging to the officials of the Government, and used for Government purposes. The common or "native" people subsist almost entirely on a fish diet — their lakes and rivers furnishing them with abundance of fish; and the whole of the summer is spent in catching and drying these for their winter provision. Several wild vegetable productions are added — roots and berries, and even the bark of trees — all of which are eaten along with the dried fish. Wild animals also furnish part of their subsistence; and it is by the skins of these — especially the sable — that the people pay their annual tax, or tribute, to the Russian Government. From animals, too, their clothing is chiefly manufactured; and many other articles used in their domestic economy.

The peninsula is rich in the fur-bearing quadrupeds, and some of these furnish the very best quality of furs that are known to commerce. The sable of Kamschatka is of a superior kind as also the many varieties of the fox. They have, besides, the wolverine and wolf, the ermine and Arctic fox, the marmot and polar-hare, and several smaller animals that yield furs of commercial value. The sea-otter is common upon the coasts of Kamschatka; and this is also an object of the chase — its skin being

13   s

among the costliest of "peltries." The great *argali*, or
wild sheep, and the reindeer, furnish them both with
flesh and skins ; but one of the chief objects of the chase
is that great quadruped for which our young hunters had
come all the way to Kamschatka, the bear. Into his
presence they would find no difficulty in introducing
themselves : for perhaps in no country in the world does
master Bruin's family muster so strongly as in this very
peninsula.

CHAPTER LII.

FISHING-BEARS.

PREVIOUS to starting forth in search of the Kam-schatkan bear, our hunters collected all the particulars they could in regard to the haunts and habits of this animal.

They learnt that there were at least two varieties known to the Kurilski and Koriac hunters. One of them was the more common kind — a brown bear, closely resembling the *ursus arctos;* and the other also a brown bear; but with a whitish list running up from the under part of his throat, and meeting like a collar over the tops of his shoulders. This latter kind was un-doubtedly the species known as the "Siberian bear" (*ursus collaris*); and which has an extensive range throughout most of the countries of Northern Asia. The native hunters alleged that the two kinds were of nearly similar habits. Both went to sleep during the winter — concealing themselves cunningly in caves and crevices among rocks, or among fallen timber, where such could be found in sufficient quantity to afford them shelter.

One remarkable habit of these bears indicates a very marked difference between them and the *ursus arctos*, with which they have been usually classed; and that is,

that they are *fishing-bears* — subsisting almost **exclusively** on fish, which they catch for themselves. During
their winter sleep, of course they eat nothing; but in
spring, as soon as they emerge from their retreats, they
at once betake themselves to the numerous streams and
lakes with which the country abounds; and roaming
along the banks of these, or wading in the water itself,
they spend the whole of their time in angling about after trout **and salmon. These fish,** thanks to their im-
mense numbers, and the shallowness of the water in
most of the lakes and streams, the bears are enabled to
catch almost at discretion. **They wade into** the water,
and getting among the shoals **of the fish as** they are passing to and fro, strike them dead with their paws. The
fish are killed as instantaneously as if impaled **upon a**
fishing-spear; and in such numbers do the bears capture
them, at certain seasons, that the captors grow dainty,
and only eat a portion of each fish! They show a
strange preference for that part which is usually considered refuse, the head, — leaving the tail, with a considerable portion of the body, untouched. The rejected
portions, however, are not lost; for another animal, still
hungrier than the bears, and less skilful **in** the piscatory
art, is at this time also in search of a meal of fish.

This creature is the Kamschatkan dog — not a wild
species, as you may suppose, but the trained sledge-dogs
of the Kamschatdales themselves; which at this season
forsake the "ostrogs," or villages, and betake themselves
to the borders of the lakes and rivers. There they remain during the whole period of summer, feeding upon
fish — which they also know how to capture — and eat-

ing up such portions as have been refused by the bears.
In fact, this is the only food which these poor dogs can
get; and, as they are not needed during the summer
season, they do not think of returning home until frost
sets in. Then strange to say, one and all of them go
voluntarily back, and surrender themselves up to their
old masters — hard taskmasters too, who not only work
them like slaves, but half starve them throughout the
whole winter. This voluntary submission to their
"yoke" has been quoted as an illustration of the high
training and faithful disposition of the Kamschatkan
dogs; but it has its origin in a far different motive than
that of mere fidelity. Their return to the snug shelter
of the *balagan* is simply an instinct of self-preservation:
for the sagacious animals well know, that in winter
the lakes and streams will be completely frozen over,
and were they to remain abroad, they would absolutely
perish either from hunger or cold. Even the wretched
winter allowance of heads and entrails of fish — the only
crumbs that fall to their share — is better than nothing
at all; which would be their portion were they to remain
abroad among the bare, snow-clad hills and valleys of
Kamschatka.

The Kamschatdales have various modes of taking the
bear. In early winter they sometimes find his track in
the snow; and then pursue him with a gun and a bear-
spear, killing him as they best can. Later still, when he
has gone to sleep in his den, he is often found — by similar indications as those which guide the Laplanders,
North American Indians, and Esquimaux — such as the
hoar caused by his breath showing over the spot, or by

their hunting-dogs scenting him out, and barking at the
entrance. The log-trap, or dead-fall, is also in use
among the Kamschatkan hunters; and the pen formed
around the mouth of the bear's cave, shutting him up,
until an entrance can be dug into it from above.

In the summer time the mode is different. Then **the**
hunter lies in ambush, with his loaded rifle — for **the**
Kamschatdale carries this weapon — in such places as
he expects the bear to pass. These are on the banks
of the streams and lakes that abound in fish; and as the
bears ramble along the edge of the water, or are even
seen swimming or wading into it, the patient hunter is
pretty sure of getting a shot. **Should he fail to bring**
down Bruin at the first fire, the game becomes uncertain; and sometimes dangerous: since the animal often
charges upon the hunter. Even though the latter may
be concealed among the long reeds and bushes, the sagacious bear, guided by the smoke and blaze of the powder
easily finds out his assailant. The hunter, however,
never fires without taking a deliberate aim. He carries
a forked stick, over which he rests his piece, and never
fires off-hand. To miss would not only endanger his
life and the loss of his game, but what is also of consequence to a Kamschatdale, the loss of his powder and
bullet — costly articles in this remote corner of the earth.
In case of missing, he has still his bear-spear and a long-
bladed knife to fall back upon; and with these he defends
himself as well as he can — though not unfrequently
Bruin proves the victor, and the hunter the victim.

There are certain times when the Siberian bears become exceedingly dangerous to approach. The season

of rut — which occurs in the latter part of the summer —
is one of these; but there is another period of danger —
which, however, does not happen every year. When the
spring chances to be late — on account of a prolonged
winter — and when the lakes and streams remain frozen
over after the bears have come forth from their hiding-
places, then "ware Bruin" is a caution which it is
prudent to observe. The fierce animals, half-famished
for want of their usual diet of fish, roam over the country
in all directions; and fearlessly approach the "ostrogs,"
roaming around the balagans and jourts in search of
something to eat. Woe to the Kamschatdale that gets
in their way at such a time — for the bear, instead of
waiting to be attacked, becomes himself the assailant;
and, as great numbers of these quadrupeds often troop
about together, of course the encounter is all the more
perilous.

It was just in such a spring that our young hunters
had arrived at Petropaulouski; and stories of numerous
bear conflicts, that had recently occurred in the neighborhood, were rife in the village; while the number of fresh
skins every day brought in by the Kurilski hunters,
showed that bears could not be otherwise than plentiful
in the country adjacent.

Guided by one of these hunters, our party set forth
upon a search. The snow still covered the ground; and,
of course, they travelled in sledges — each having one
to himself, drawn by five dogs, as is the custom of the
country. The dogs are harnessed two and two abreast,
with the odd one in front. Each has his collar of bear-
skin, with a leather thong for a trace; and five of them

are sufficient to draw the little sledge with a man in it. The sledge, called *saunka*, is less than four feet long; and, being made of the lightest birch wood, is of very little weight.

A curved stick, called the *oschtol* — with an iron point, and little bells at the other end — is used to direct the dogs; and, urged on by this and by well-known exclamations of their driver, they will go at a speed of many miles an hour.

In this slight vehicle, hills, valleys, lakes, and rivers are crossed, without such a thing as a road being thought of; and when the dogs are good, and have been well cared for, an immense distance may be passed over in a day.

In less than an hour after their departure from Petro-paulouski, our hunters had entered amid the wildest scenery — where not the slightest sign of either cultivation or human habitation was to be seen, and where at any moment they might expect to come in sight of their great game.

CHAPTER LIII.

DOG-DRIVING.

THE guide was conducting them to a stream that ran into the bay some ten or twelve miles from the "ostrog." On that stream, he said, they would be pretty certain to find a bear, if not several: since at a place he knew of the water was not frozen, and the bears might be there trying to catch fish. When questioned as to why this particular stream was not frozen like the others, he said that some distance up it there were warm springs — a phenomenon of frequent occurrence in the peninsula of Kamschatka — that these springs supplied most of the water of the stream; and that for several hundred yards below where they gushed forth, the river was kept open by their warmth during the severest winters. Not throughout its whole course, however. Farther down, where the water became cool, it froze as in other streams; and that this was the case, was evident to our hunters, who had entered the mouth of the rivers from the icy surface of the bay, and were gliding in their sledges up its frozen channel.

After having gone three or four miles up this ice-bound stream, which ran through a narrow valley with steep, sloping sides, the guide warned our hunters that

13 *

they were close to the place where the water would be
found open. At this point a low ridge ran transversely across the 'valley — through which the stream had,
in process of time, cut a channel; but the ridge oc-
casioned a dam or lake of some half-dozen acres in superficial extent, which lay just above it. The dam itself was rarely frozen over; and it was by the water
remaining in it, or flowing sluggishly through it — and
thus giving it time to cool — that the stream immediately below got frozen over.

The lake lay just on the other side of the ridge,
and was now only hidden from their view by the rise
of the ground. If not frozen over, as the guide conjectured, there was likely to be a bear roaming around
its edge ; and therefore they resolved to observe caution
in approaching it.

The sledges were to be taken no further. Our hunters had learnt how to manage both dog-sledges and dogs.
Their experience in Finland, as well as in the countries
of the Hudson's Bay territory, had taught them that;
and made them skilful in the handling of these animals
— else they would have made but poor work in travelling
as they did now. In fact, they could not have managed
at all : since it requires a great deal of training to be
able to drive a dog-sledge. This, however, they had received — both the boys and Pouchskin — and fortunate
it had been so; for very shortly after they were placed
in a predicament, in which their lives depended on their
skill as sledge drivers.

The dogs were left under cover of the ridge, near
the bottom of the little slope ; a sign was given to them

to keep their places — which these well-trained creatures perfectly comprehended ; and the hunters — the Kurilski with the rest — holding their guns in readiness, ascended towards the summit of the slope.

There was no cover, except what was afforded by the inequality of the ground. There were no trees in the valley — only stunted bushes, not half the height of of a man's body, and these nearly buried to their tops in the snow. A few, however, appeared growing along the crest of the ridge.

The hunters crawled up to these on all-fours, and peeped cautiously through their branches.

It was the impatient Ivan that looked first; and what he saw so surprised him as almost to deprive him of the power of speech! Indeed, he was not able to explain what he saw — till the other three had got forward, and became equally eyewitnesses of the spectacle that had astonished him.

As the guide had conjectured, the lake was not frozen. There was some loose snow floating over its surface; but most of the water was open; and the stream that flowed slowly in on the opposite side was quite clear of either ice or snow.

The guide had also predicted hypothetically that they might see a bear — perhaps two. It had not occurred to this man of moderate pretensions that they might see *twelve* — and yet no less than twelve bears were in sight !

Yes, twelve bears — they were as easily counted as oxen — were around the shores of this secluded lake, and on the banks of the little stream that ran into it — all

within five hundred yards of each other. Indeed, it
would have been easy to have mistaken them for a herd
of brown heifers or oxen ; had it not been for the various attitudes in which they were seen : some upon all-
fours — some standing erect, like human beings, or
squatted on their hams like gigantic squirrels — others in
the water, their bodies half submerged — others swimming about, their backs and heads only visible above the
surface; and still others prowling leisurely along the
banks, or over the strip of level meadow-land that bor-
dered the lake.

Such a sight our bear-hunters had never witnessed
before, and might never witness again, in any other country, save Kamschatka itself. There it is by no means
uncommon ; and twenty bears instead of twelve have
been often seen in a single drove — at that season when
they descend from their mountain retreats to their favorite
fishing-grounds upon the lakes and streams.

Our hunters were perplexed by so unexpected a sight ;
and for some moments unresolved as to how they should
act. Fortunately, the bushes already mentioned served
to conceal them from the bears ; and the wind was blow-
ing towards the hunters — otherwise the bears, who are
keen of scent, would soon have discovered their presence.
As it was, not one of them — though several were close
to the ridge — seemed to have any suspicion that an
enemy was so near. The huge quadrupeds appeared
to be too busy about their own affairs — endeavoring
to capture the fish — some of them greedily devouring
those they had already taken, and others wandering rest-
lessly about, or eagerly observing the movements of the

fish in the water. One and all of them looked fierce and famished, their bodies showing gaunt and flaky, and their enormous limbs having a lank, angular appearance, that gave them a still greater resemblance to heifers — only heifers that had been half starved !

CHAPTER LIV.

A SLEDGE-CHASE.

I HAVE said that our hunters were for some time irresolute about how to act. The Kurilski was inclined to withdraw from the spot and leave the bears alone; and this of course was his advice to the others. He said there might be danger in disturbing them — so many clustered together, and in such a mood as they appeared to be. He had known them to attack a large party of men under such circumstances, and give chase to them. They might do the same now.

Our hunters, however, did not give full credit to this story of their guide — thinking it might have its origin in the fears of the Kurilski, whom they knew to be of a timid race; and therefore they determined not to back out. The chance was too tempting to be surrendered for so slight a reason, and without a struggle. There were several bears within easy shot of the ground where they were kneeling!

It would never do to let such an opportunity pass. They might not meet with so good a chance again; or, at all events, they might be delayed a good long time before another would turn up; and a residence in Petro-paulouski, even in the "isba" of the governor — who was himself only a sergeant of Cossacks, and his dwelling a mere hut — was not so pleasant as that they should wish to prolong it. They had now been a great while journeying through countries covered with frost and snow ; and they were longing to reach those tropical isles — famed for their spices and their loveliness — which were to be the next stage in their grand tour round the globe.

Influenced by these thoughts, then, they resolved to run all hazard, and try a shot at the bears.

The Kurilski, seeing them determined, gave in ; and, joining his gun to theirs, a volley of four shots was simultaneously discharged through the bushes.

Two bears were seen to drop over and lie kicking upon the snow ; but whether they continued their kicking for any considerable length of time, was a question about which our hunters could give no definite information. They did not stay to see : for the moment the smoke had cleared off, they saw the whole gang of bears in motion, and rushing towards them from all sides of the lake. The shrill, fierce screaming of the animals, and the hurried pace in which they were making towards the ridge, declared their intentions. They were charging forward to the attack.

The hunters saw this at a glance ; and thought only of retreating. But whither could they fly ? There were no trees ; and if there had been, the bears could have climbed them even better than themselves. There were steep, rocky cliffs on both sides of the ravine ; but these would afford them no security — even had their ice-coated slope permitted of their being scaled. But it did not, and if it had, the bears could have scaled the rocks too !

Our Russian hunters were in a complete state of perplexity, and perhaps would not have known how to save themselves, had it not been for their Kurilski comrade. He, however, had conceived an idea — or, rather, had drawn it from old experience; and just at this moment he rushed down the slope, as he did so calling to the others to take to their sledges, and warning them that it was their only chance of escape.

Of course none of them thought of disputing his advice, or even calling it in question; but one and all of them yielded obedience on the instant. Without saying a word, each rushed to his sledge, leaped upon the runners, seated himself in double quick time upon the little crescent-like cradle, seized the "ribbons," and straightened his team to the road.

Had the dogs not been well trained, and their drivers equally well used to the managemement of a sledge, their peril would have been extreme. As it was — though all came into their places in good style, and without confusion — they had not a second to spare. The bears were already galloping down the slope; and as the last sledge — which was Pouchskin's — moved off from the bottom of the ridge, the foremost of the roaring pursuers had got within less than six yards of it !

It was now a trial of speed between bears and sledge-dogs — for the latter knew that they were in as much danger as their masters; and needed neither the exclamation *Ah!* nor the *oschtol* to urge them forward. On swept they over the frozen crust, as fast as they could go — handling their limbs and claws with the nimbleness peculiar to their race.

The bears followed in a sort of lumbering gallop; yet, notwithstanding their uncouth movements, they kept, for a long time, close in the rear of the fugitives.

Fortunately they did not possess the speed of the canine race; and at length — seeing that they were being distanced — one after another gave up the chase, and commenced returning towards the lake, slowly, and with apparent reluctance.

Just at this crisis an accident occurred to Pouchskin — or rather Pouchskin committed a mistake — which, had it been made five minutes sooner, would most assuredly have cost him his life. The mistake which Pouchskin made, was to drop the iron end of his "*oschtol*" on the snowy crust between his sledge and the two dogs nearest to it — the "wheelers" as we may call them. The effect of this, with Kamsckatkan sledge-dogs, is to cause the whole team to halt; and so acted the dogs that Pouchskin was driving — all five suddenly coming to a dead stop! Pouchskin endeavored to urge them forward — crying out the usual signal, *Ha*; but, in his anxious eagerness, Pouchskin placed the accent after the vowel, instead of before it; and instead of *Ha!* his exclamation sounded *Ah!* The latter being the command for the dogs to halt, of course only kept them steady in their places; and they stood without offering to move a leg. By good fortune, the bears had already given up the pursuit, and were not witnesses of this interruption: otherwise it would have gone ill with the ex-grenadier.

T

In due time the dogs were once more started ; and Pouchskin — putting them to their highest rate of speed — soon overtook the sledge-train ; which did not come to a halt until a good mile of snow-covered country was between it and the bears.

The hunters only paused then, for a short while, to breathe their panting dogs ; and this done, they resumed their seats on the sledges, and continued on to the ostrog — without a thought of going back after the bears.

They had no intention, however, of giving them up entirely. They only drove home to the village — in order to get assistance ; and, as soon as their report was delivered, all the men of the settlement — Cossacks, Kurilskis, and half-breeds — turned out armed to the teeth for a grand battue, and proceeded towards the lake with the governor himself at their head.

The bears were still upon the ground — both the living and the dead — for it was now seen that two of their number had fallen to the shots of our hunters — and upon the former a general fusillade was at once opened, which ended in their complete discomfiture. Five more of them were killed upon the spot ; and several others that took to flight were tracked through the snow, and destroyed in their hiding-places. For a week after, there was very little fish eaten in the ostrog of Petropaulouski — which, for a long period previous to that time, had not witnessed such a carnival.

Of course our Russian hunters came in for their share of the trophies ; and, choosing the skin of one of the bears they had themselves shot, they left it with the Governor, to be forwarded *viâ* Okhotsk and Yakoutsk, to the distant capital of St. Petersburg. Shortly after the fur-ship carried them to Canton, — whence they might expect to find a passage in a Chinese trading-vessel to the grand island of Borneo.

CHAPTER LV.

THE SUN BEARS.

THERE are colonies of Chinese settled in different parts of Borneo — whose principal business there is the working of gold and antimony mines. These Chinese colonial settlements — along with numerous others throughout the Oriental islands — are under the protection and direction of a great Mercantile Company called *Kung Li* — somewhat resembling our own East India Company. In Borneo, the head-quarters of this commercial association of the Chinese, is the port and river of Sambos, on the western coast; though they have many other settlements in different parts of the island. Of course, between these colonies and Canton there is a regular traffic; and our travellers found no difficulty in proceeding to Borneo in a Chinese junk which traded direct from Canton to Sambos. At Sambos there is also a Dutch settlement, or "factory," belonging to the Dutch East India Company; and this Company has also two other stations in the island — all, however, occupying a territory of limited extent, compared with the large surface of the island itself. No other European settlements exist in Borneo, if we except an English "agency" lately established at the little island of Labuan; and a settlement at Sarawak, under an English adventurer, who styles himself "Rajah Brooke."

The "rajah" rests his claim to the title and territory of Sarawak on a grant from the Sultan of Borneo (Bruni) ; and the *quid pro quo* which he professes to have given, was the having assisted the said Sultan in putting down the "Dyak pirates!" This is the pretence hitherto put forth to the British public ; but on a closer inquiry into the facts of this transaction, the story assumes quite a different color ; and it would rather appear that, instead of assisting to put down piracy in the Bornean waters, the first act of the philanthropic Englishman was to assist the Malay Sultan in enslaving several tribes of inoffensive Dyaks, and forcing them to work without pay in the mines of antimony ! This appears to have been the nature of the services that purchased Sarawak. It was, in fact, aiding the pirates, instead of putting them down : since the Bornean Sultan was himself the actual patron and protector of these sea robbers, instead of being their enemy !

The patriot and statesman, Hume, endeavored to procure an inquiry into these acts of Oriental filibusterism ; but the underhand influence of an unprincipled Administration, backed by an interested commercial clamor, was too strong for him ; and the shameful usurpation has been justified.

Notwithstanding that Europeans have been settled for hundreds of years in the islands of the Indian Archipelago — ruling them, as we may almost say — it is astonishing how little is yet known of the great island of Borneo. Only its coasts have been traced, and these very imperfectly. The Dutch have made one or two expeditions into the interior; but much knowledge need not be expected from such trading hucksters as they. Their energies in the East have been expended throughout a period of two centuries, with no other apparent object than to promote dissension, wherever it was possible; and to annihilate every spark of freedom or nobility among the races who have had the misfortune to come in contact with them.

Notwithstanding their opportunities, they have done little to add to our knowledge of Borneo — which was about as well known a hundred years ago as it is at the present hour. Never was a subject more ripe for illustration than this magnificent island. It courts a monograph — such as has been given to Sumatra by Marsden, by Tennant to Ceylon, and to Java by Sir Stamford Raffles. Perhaps some one of my young readers may become the author of that monograph.

Teeming with the most gorgeous forms of tropical life — so rich in *fauna* and *flora* that it might be almost regarded as a great zoölogical and botanical garden combined — it will well repay the scientific explorer, who may scarce find such another field on the face of the earth.

Our young hunters, in contemplating the grand tropical scenery of Borneo, were filled with admiration. The *sylva* was quite equal to anything they had witnessed on the Amazon; while the *fauna* — especially in quadrupeds and *quadrumana* — was far richer.

To one quadruped was their attention more especially directed; and I need hardly say that this was the Bornean bear — by far the most beautiful animal of the whole Bruin family. The Bornean bear is also the smallest of the family — in size, being even less than his near congener, the Malayan bear; though resembling the latter in many particulars. His fur is a jet black, with a muzzle of an orange-yellow color, and a disc of still deeper orange upon the breast, bearing a certain resemblance to the figure of a heart. The hair is thickly and evenly set over his whole body — presenting the same uniform surface which characterizes the black bear of North America, the two species of South America, and also his Malayan cousin — who inhabits the neighboring islands of Sumatra and Java. For the latter, indeed, he is often taken; and many naturalists consider them as one species — though this is certainly an error. The Bornean bear is not only much less in bulk; but the deep orange-color on his breast offers a permanent mark of distinction. In the Malayan bear there is also a marking on the breast; but it is of half-moon shape and whitish color. Besides, the color of the muzzle in the latter species is only *yellowish*, not *yellow;* and the animal altogether is far from being so handsome as the bear of Borneo.

Dr. Horsfield, who had good opportunities of observing them both, has pointed out other essential characteristics, which prove conclusively that they are separate species; but the Doctor, guided by his love for generic distinctions, could not rest satisfied, without further ornamenting his task — by constituting for them a new genus, under the title of *Helarctos.* There is no reason whatever for this inundation of generic names. It has

served no good purpose; but, on the contrary, renders
the study of natural history more complicated and ob-
scure; and to no family of animals do these remarks
more pointedly apply than to that of the bears. So
similar are all these quadrupeds to one another — so perfect is the *family likeness* between them — that to separate them into different genera is a mere pedantic conceit
of the anatomists. There are about a dozen species in
all; and the systematic naturalists — who do not even
admit that number — have formed for the bears nearly
as many genera as there are species, — among which
may be mentioned the ridiculous titles of *Proehilus*,
Melursus, *Helarctos*, and the like.

The Bornean bear is as much a true species of *ursus*
as either the brown bear of Europe, the black bear of
North America, or the black bears of the Cordilleras;
and, indeed, to these last his habits assimilate him very
closely — being, like them, a vegetarian in his diet, and
a great lover of sweets.

Of his *penchant* for honey our young hunters had
proof: for, it was while actually engaged in plundering
a hive they first saw the Bornean bear. They were at
the same time successful in effecting his capture — which
is now to be described.

CHAPTER LVI.

THE TALL TAPANG.

On their arrival at Sambos, our young hunters, according to their usual custom, procured a native guide to direct them to the haunts of their game. In this case it was a Dyak who became their conductor — one of those who follow the business of bee-hunters ; and who, from the very nature of their calling, are often brought into contact with the bears as well as the bees.

Under the direction of the Dyak, our hunters made an excursion to a range of wooded hills, not far from Sambos, where the sun bear was known to exist in great numbers ; and where one was likely to be found almost at any time.

As they were passing through the woods, they observed a very singular species of tree — indeed, many species, that might be styled singular ; but one pre-eminently so, that strongly arrested their attention. These trees did not grow in any great numbers together ; but only two or three in one place ; and more generally they stood singly — apart from any of their own kind, and surrounded by other trees of the forest. But though surrounded by other sorts, they were overtopped by none. On the contrary, their own tops rose above all the others

14

to a vast height; and what was most singular, they did
not put forth a branch from their trunks until the latter
had shot up to some feet above the "spray" of the
surrounding forest. It was this peculiarity that had
drawn the attention of our hunters. They might not
have noticed it, had they kept on under the trees; but,
on crossing a slight eminence — where the ground was
open — they chanced to get a view of a number of these
tall trees, and saw that they towered to a vast height,
above all the others.

Even their tops had the appearance of tall trees, standing thinly over the ground — the ground itself being
neither more nor less than the contiguous heads of the
other trees, that formed the forest. Had this forest been
a low jungle, there would have been nothing extraordinary
in what they saw; but our hunters had already observed
that it was a true forest of grand trees — most of them
a hundred feet in height. As the trees which had attracted their admiration rose full fifty feet above the tops
of the others, it may be imagined what tall individuals
they were. They were slender, too, in proportion to
their height; and these stems rising two hundred feet,
without a single offshoot or branch upon them, gave the
trees the appearance of being still taller than they actually were — just as a thin, clean spar, set upright, looks
much taller than a hill or a house of the same elevation.

We have said that there were no branches for the first
hundred feet or so up the stem. Beyond that there
were many and large limbs; which, diverging only
slightly, and in a fastigiate manner, carried the tree nearly
as much higher. These branches were regularly set;

and covered with small, light-green leaves, forming a beautiful, round head.

The bark of this tree was white, and by piercing it with a knife, our hunters perceived that it was soft and milky. The wood, too, for some inches below the periphery was so spongy, that the blade of the knife penetrated into it almost as easily as into the stalk of a cabbage.

The wood near the bark was of a white color. Inwards it became harder; and had they been able to reach the heart, they would have found it very hard, and of a dark chocolate color. On exposure to the air, this heart-wood turns black as ebony; and is used for similar purposes by the native Dyaks and Malays, who manufacture from it bracelets and other *bijouterie*.

On asking their Dyak guide the name of this remarkable tree, he said it was called the *tapang*. This, however, gave no information regarding its species; but Alexis, shortly after, in passing under one, observed some flowers that had fallen from its top; and having examined one of these, pronounced the tree a species of *ficus* — a very common genus in the islands of the Indian Archipelago.

If our young hunters were filled with admiration at sight of this beautiful tree itself, they shortly after observed something that changed their admiration into wonder. On advancing towards one of the tapangs, they were struck with a singular serrated appearance that showed along the edge of its trunk — from the ground up to the base of its branching head. It looked as if a tall ladder was laid edgeways along the trunk of the tree — one side of it hidden under the bark! On drawing

nearer, this appearance was explained. A ladder in
reality it was ; but one of rare construction ; and which
could not have been removed from the tree, without taking it entirely to pieces. On closer examination, this
ladder proved to be a series of bamboo spikes — driven
into the soft trunk in a slightly slanting direction, and
about two feet apart, one above the other. The spikes
themselves forming the rounds, were each about a foot in
length ; and held firmly in their places by a bamboo rail
— to which their outer ends were attached by means of
thin strips of rattan. This rail extended the whole way
from the ground to the commencement of the branches.

It was evident that this extemporized ladder had been
constructed for the purpose of climbing the tree, but with
what object ? Upon this head their Dyak guide was the
very man to enlighten them : since it was he himself
who had made the ladder. The construction of such ladders, and afterwards the climbing of them, were the most
essential branches of his calling — which, as already
stated, was that of a bee-hunter. His account of the
matter was as follows. A large wasp-like bee, which is
called *lanyeh*, builds its nests upon these tall tapangs.
The nest consists of an accumulation of pale, yellowish
wax — which the bees attach to the under-side of the
thick branches, so that these may shelter the hive *from*
the rain. To reach these nests, the bamboo ladder is constructed, and the ascent is made — not for the purpose
of obtaining the honey alone — but more on account of
the wax, out of which the combs are formed. The
lanyeh being as much *wasp* as *bee*, produces a very small
quantity of honey ; and that, too, of inferior quality ; but

the wax is a valuable article, and of this several dollars' worth may be procured from a single hive.

It is dearly earned money — very dearly earned, indeed; but the poor Dyak bee-hunter follows the calling from motives not easily understood — since almost any other would afford him a living, with less labor and certainly with less *pain*. Pain, indeed! He never succeeds in plundering the store of the *lanyeh*, without being severely stung by the insects; and though their sting is quite as painful as that of the common wasp, experience seems to have rendered the Dyak almost indifferent to it. He ascends the flimsy ladder without fear — carrying a blazing torch in his hand, and a cane-basket on his back. By means of the torch, he ejects the bees from their aerial domiciles; and, then having torn their combs from the branches, he deposits them in his basket — the incensed insects all the while buzzing around his ears, and inflicting numerous wounds over his face and throat, as well as upon his naked arms! Very often he returns to the ground with his head swollen to twice the size it was previous to his going up! Not a very pleasant profession is that of a Bornean bee-hunter!

CHAPTER LVII.

THE BRUANG.

As the party proceeded onward, they observed several other tapang-trees, with ladders attached to them; and at the bottom of one of these — which was the tallest they had yet seen — the guide made a halt.

Taking off his *kris*, and throwing to the ground an axe, which he had brought along, he commenced ascending the tree.

Our hunters inquired his object. They knew it could not be either honey or wax. There had been a bees'-nest upon this tree — as the ladder told — but that had been removed long ago; and there now appeared nothing among the branches that should make it worth while to climb up to them. The answer of the bee-hunter explained his purpose. He was merely ascending to have a lookout over the forest — which in that neighborhood could not be obtained by any other means than by the climbing of a *tapang*.

It was fearful to watch the man ascending to such a dizzy height, and with such a flimsy, uncertain support beneath his feet. It reminded them of what they had seen at the Palombière of the Pyrenees.

The Dyak soon reached the top of the ladder; and

for some ten minutes or more clung there — screwing his head around, and appearing to examine the forest on all sides. At length his head rested steadily upon his shoulders; and his gaze appeared to be fixed in one particular direction. He was too distant for the party at the bottom of the tree to note the expression upon his countenance; but his attitude told them that he had made some discovery.

Shortly after he came down; and reported this discovery in laconic phrase, simply saying: —

" *Bruang* — see him ! "

The hunters knew that " bruang " was the Malayan name for bear; and the coincidence of this word with the *sobriquet* " Bruin " had already led them to indulge in the speculation, as to whether the latter might not have originally come from the East.

They did not stay to think of it then: for the guide, on regaining *terra firma*, at once started off — telling them to follow him.

After going rapidly about a quarter of a mile through the woods, the Dyak began to advance more cautiously — carefully examining each of the trunks of the *tapangs* that stood thinly scattered among the other trees.

At one of these he was seen to make an abrupt halt, at the same instant turning his face upward. The young hunters, who were close behind him, could see that there were scratches upon the soft, succulent bark, as if caused by the claws of some animal; but, almost as soon as they had made the observation, their eyes were directed to the animal itself.

Away up on the tall tapang — just where its lowest

limbs parted from the main stem — a black body could be distinguished. At such a distance it appeared not bigger than a squirrel; but, for all that, it was a Bornean bear; and the spot of vivid orange upon its breast could be seen shining like a coal of fire. Close by its snout a whitish mass appeared attached under the branches. This was the waxen domicile of the *lanyeh* bees; and a slight mist-like cloud, which hung over the place, was the swarm itself — no doubt engaged in angry conflict with the plunderer of their hive.

The little **bear was** too busy in the enjoyment of his luscious meal — that is, if the stings of the *lanyehs* allowed him to enjoy it — **to look below; and** for some minutes the hunters stood regarding him, without making a movement.

Satisfied with their inspection, they were at length preparing to fire at him; when they were hindered by the Dyak — who, making signs to them to be silent, drew them all back from **the** tree.

When out of sight of the bear, he counselled them to adopt a different plan. He said — what was true enough — that at such a height they might miss the bear; or, even if they should hit him, a bullet would scarce bring him down — unless it should strike him in a vital part. In the contingency of their missing, or only slightly wounding him, the animal would at once ascend further up into the tapang; and, hidden behind the leaves and branches, might defy them. He would there remain till hunger should force him down; and, since he was just in the act **of** having his meal, and had, no doubt, been eating from the time he was first espied — or longer, perhaps — he would be in a condition to stay in the tree, until their patience should be more than exhausted.

True, they might fell the tree; they had an axe, and could soon cut the tree down — as the wood was soft; but the Dyak alleged that the bruang in such cases usually contrives to escape. The tapang rarely falls all the way, but only upon the tops of the trees that stand thickly round; and as the Bornean bear can climb and cling like a monkey, he is never shaken out of the branches, but springs from them into some other tree — among the thick leaves of which he may conceal himself; or, by getting to the ground, manage to steal off.

His advice, therefore, was, that the hunters should conceal themselves behind the trunks of the surrounding trees; and, observing silence, wait till the bruang had finished his mellifluous repast, and feel inclined to come down. The Dyak said he would make his descent stern foremost; and, if they acted cautiously, they might have him at their mercy, and almost at the muzzles of their guns.

There was only one of the three who was not agreeable to this plan; and that was the impatient Ivan; but, overruled by the advice of his brother, he also gave his consent to it.

The three now took their respective stands behind three trees — that formed a sort of triangle around the tapang; and the guide, who had no gun, placed himself apart — holding his kris in readiness to finish off the bear, should the animal be only wounded.

There was no danger to be dreaded from the encounter. The little bear of Borneo is only dangerous to the bees and white ants — or other insects — which he is accustomed to lick up with his long tongue. The human hunter has nothing to fear from him, any more than from a timid deer — though he will scratch, and growl, and bite, if too closely approached.

It was just as the Dyak had predicted. The bruang, having finished his meal, was seen coming down the tree tail foremost; and in this way would no doubt have continued on to the ground ; but, before he had got half-way down the trunk, Ivan's impatience got the better of him ; and the loud bang of his fowling-piece filled the forest with its echoes. Of course it was a bullet that Ivan had fired ; and it appeared that he had missed. It was of little use firing also his shot barrel, though he did so immediately after.

The effect of his shots was to frighten the bruang back up the tree ; and at the first report he commenced ascending. Almost as rapidly as a cat he swarmed upward ; and for a moment the chances of losing him appeared as two to one. But Alexis, who had been watching the restless movements of his brother, had prepared himself for such an issue ; and, waiting till the bruang made a pause just under the branches, he fired his rifle with deadlier aim. The bear, in clutching to one of the limbs, had extended his body outward, and this gave the rifleman the chance of aiming at his head. The bullet must have told : for the bear, instead of ascending higher, was seen hanging down from the limb, as if he was clinging to it with enfeebled strength.

At this moment the cannon-like report of Pouchskin's fusil filled the woods with its booming echoes; and Bruin,

suddenly relaxing his grasp, came bump down among
the hunters — missing Pouchskin by about the eighth
part of an inch! Lucky for the old grenadier there was
even this much of a miss. **It was** as good as a mile to
him. Had the bear's body descended upon his shoulders,
falling from such a height, it would have flattened him out
as dead as the bear was himself; and Pouchskin, perceiving the danger from which he had so narrowly escaped,
looked as perplexed and miserable as if some great misfortune had actually befallen him!

CHAPTER LVIII.

THE CABBAGE-EATER.

OUR heroes now, having accomplished their mission to Borneo, were about to cross over to the island of Sumatra; in which — as **well as in** Java, or upon the mainland of Malacca — they would **find the** other sun bear, known as the *ursus malayanus;* but previous to their departure from Sambos, they obtained information that led them to believe that this species also inhabited the island of Borneo. It was more rarely met with than the orange-breasted variety; but the natives, generally better guides than the anatomists in the matter of specific distinctions, stoutly maintained that there were two kinds; and the Dyak bee-hunter — whose interest had been secured by the ample reward already bestowed upon him — promised them, that if they would go with him to a certain district of country, he would show them the larger species of bruang. From the man's description of it Alexis easily recognized the *ursus malayanus* — the species they had killed being the *ursus euryspilus.*

Indeed, had there been any doubt about this matter, it would have been set at rest, by what our travellers saw in the streets of Sambos. There both species were exhibited by the itinerant jugglers — for both the sun bears can be easily tamed and trained — and these men stated that they had procured the "big bruang," in the woods of Borneo.

Since, then, he was there to be found, why go to Sumatra in search of him? They had still travelling enough before them; and they were beginning to get tired of it. It was natural that — after so long an absence and the endurance of so many perils and hardships — they should be longing for home, and the comforts of that fine palace on the banks of the Neva.

They resolved, therefore, to accompany the Dyak guide on a new expedition.

They were a whole day upon the journey; and just before nightfall reached the place, where the man expected to fall in with the big bruangs. Of course, they could not commence their search before morning. They halted, therefore, and formed camp — their Dyak guide erecting a bamboo hut in less than an hour, and thatching it over with the huge leaves of the wild *musaceæ*.

The place where they had halted was in the midst of a magnificent grove, or rather a forest, of palms; of that kind called *nibong* by the natives, which is a species of the genus *arenga*. It is one of the "cabbage" palms; that is, its young leaves before expanding are eaten by the natives as a vegetable — after the manner in which Europeans use cabbage. They are of a delicate whiteness, with a sweet nutty flavor; and, in point of excellence, are even superior to those of the cocoa-nut, or even the West India cabbage-palm *areca oleracea*. But the nibong is put by the Borneans and other natives of the Indian Archipelago to a great variety of uses. Its round

stem is employed as uprights and rafters for their houses.
Split into laths, it serves for the flooring. Sugar can be
obtained from the saccharine juice of its spadix, which
also ferments into an intoxicating beverage; and sago
exists in abundance within the trunk. Pens and arrows
for blow-guns are also made from the midribs of the side
leaves; and, in fact, the *arenga saccharifera*, like many
other palms, serves for an endless variety of purposes.

Alexis was greatly interested by the appearance of this
beautiful tree; but it was too late when they arrived on
the ground for him to have an opportunity of examining
it. The half-hour before darkness had been occupied in
the construction of the hut — in which all hands had
borne part.

Early in the morning, Alexis — still curious about the
arenga-trees — and desirous of ascertaining to **what**
genus of palms they belonged — strayed off among them
in hopes of procuring a flower. The others remained
by the hut, preparing breakfast.

Alexis saw none of the trees in flower, their great
spathes being yet unfolded; but, hoping to find some one
more forward than the rest, he kept on for a considerable
distance through the forest.

As he was walking leisurely along, his eyes at intervals
turned upward to the fronds of the palms, he saw that
one of the trunks directly in front of him was in motion.
He stopped and listened. He heard a sound as of something in the act of being rent, just as if some one was
plucking leaves from the trees. The sound proceeded
from the one that was in motion; but it was only its trunk
that he saw; and whatever was causing the noise and

the movement appeared to be up among the great fronds at its crown.

Alexis regretted that he had left his gun behind him. He had no other weapon with him but his knife. Not that he was afraid: for the animal could not be an elephant in the top of a palm-tree, nor a rhinoceros; and these were the only quadrupeds that need be greatly dreaded in a Bornean forest: since the royal tiger, though common enough both in Java and Sumatra, is not an inhabitant of Borneo.

It was not fear that caused him to regret having left his gun behind him; but simply that he should lose the chance of shooting some animal — perhaps a rare one. That it was a large one he could tell by the movement of the tree: since no squirrel or small quadruped could have caused the stout trunk of the palm to vibrate in such a violent manner.

I need not say how the regret of the young hunter was increased, when he approached the tree, and looking up, saw what the animal really was — a bear, and that bear the true *ursus malayanus!* Yes, there was he, with his black body, yellowish muzzle, and white half-moon upon his breast — busy gorging himself upon the tender leaflets of the arenga — whose white fragments, constantly dropping from his jaws, strewed the ground at the bottom of the tree.

Alexis now remembered that this was a well-known habit of the Malayan bear — whose favorite food is the "cabbage" of palm-trees, and who often extends his depredations to the cocoa plantations, destroying hundreds of trees before he can be detected and destroyed himself.

Of course this wild arenga wood — furnishing the bear
with as much " cabbage " as he might require — was just
the place for him ; and Alexis now understood the reason
why the Dyak had conducted them thither.

As the naturalist knew that this kind of bear was more
rare than the other species — that is, in Borneo — he
now more than ever felt chagrin at not having his gun
with him. To attempt attacking the animal with his
knife would have been absurd, as well as dangerous —
for the Malayan bear can maintain a better fight than
his Bornean brother.

But, indeed, even had Alexis desired it, there would
have been no chance to reach the animal with his
knife — unless the hunter should himself climb up the
palm ; and that was more than he either dared or could.

Of course the bear had long ere this perceived his
enemy at the foot of the tree; and, uttering a series of
low, querulous cries, had desisted from his cabbage eating,
and placed himself in an attitude of defence. It was
evident from the position he had assumed, that he had no
design of coming down, so long as the hunter remained
at the bottom of the tree; nor did the latter desire him
to do so. On the contrary, he struck the tree with a
stick, and made several other demonstrations, with the
design to hinder the bear from attempting a descent.
But the animal did not even meditate such a thing.
Though the palm was not one of the highest, it was tall
enough to keep him out of the reach of any weapon the
hunter could lay hands upon ; and the bear, seemingly
conscious of this fact, kept his perch with a confident air
— that showed he had no intention of changing his
secure position.

Alexis now began to reflect about what he should do. If he could make the others hear him, that would answer every purpose. Of course they would come up, bringing with them their guns. This was the most promising plan; and Alexis hastened to put it into execution, by hallooing at the top of his voice. But, after he had shouted for nearly ten minutes, and waited for ten more, no response was given; nor did any one make an appearance upon the ground.

Once more Alexis raised his voice, and shouted till the woods rang with echoes. But these echoes were all the reply he could get to his calls.

It was evident he had unconsciously strayed far from the camp, and quite out of earshot of his companions!

What was to be done? If he should go back to the others, to bring them and also his gun, the bear would in all probability seize the opportunity to descend from the tree and take himself off. In that case he would most certainly escape: since there would be no chance of tracking him through such a wood. On the other hand, Alexis need not remain where he was. He might stay there till doomsday, before Bruin would condescend to come down; and even should he do so, what chance would there be of effecting his capture?

While reflecting thus, a happy idea occurred to the young hunter; and he was seen all at once to step a pace or two back, and place himself behind the broad leaves of a wild *pisang*, where he was hidden from the eyes of the bear.

As the morning was a little **raw** he had his cloak around him; and this he instantly stripped off. He had already in his hands the stout long stick — with which **he** had been hammering upon the palm — and this he now sharpened at one end with his knife. On the other end he placed his cap, and beneath it his cloak, folding the latter around **the** stick, and tying it on in such a fashion as to make of it a rude representation of the human form.

When he had got the "dummy" rigged out to his satisfaction, **he** reached cautiously forward — still keeping **the** fronds of the pisang between himself and the bear. In this position, he held the "scarecrow" out at the full length of his arm; **and,** giving the stick a punch, set it erect in the ground. The bruang, from **his** elevated perch on the tree, could not fail to see the object — though the hunter himself was still concealed by the huge leaves that drooped over his head. Alexis, now cautiously, and without making the slightest noise, stole away from the spot. **When** he believed himself well out of hearing of **the** bear, he quickened his pace, and retraced **his** steps to the camp.

It was but the work of a minute for all hands **to arm** themselves and set out; and in ten minutes' time they arrived **at the** bottom of the *arenga,* and had the gratification of finding that the *ruse* of Alexis had proved successful.

The bruang was still crouching upon the crown of the palm; but he did not stay there much longer, for a volley fired at his white breast toppled him over from his perch; and he fell to the bottom of the tree as dead as a stone.

The Dyak was rather chagrined that he had not himself discovered the game; but, on ascertaining that he would receive the promised bounty all the same, he soon got the better of his regrets.

Our hunters being on the ground, were determined to make a day of it; and after breakfast continued their hunt — which resulted in their finding and killing, not only another *bruang,* but a *rimau dahan,* or " clouded tiger" (*felis macrocelus*) : the most beautiful of all feline animals, and whose skin they intended should be one of the trophies to be mounted in the museum of the palace Grodonoff.

This hunt ended their adventures in the Oriental Archipelago; and from Sambos they proceeded direct through the straits of Malacca, and up the Bay of Bengal to the great city of Calcutta.

CHAPTER LIX.

THE SLOTH BEAR.

En route for the grand mountains of Imaus — the stupendous chain of the Himalayas!

There our hunters expected to find no less than three species of bears — each distinct from the others in outline of form, in aspect, in certain habits, and even in habitat; for although all three exist in the Himalayas, each has its own zone of altitude, in which it ranges almost exclusively. These three bears are, the "sloth bear" (*ursus labiatus*), the "Thibet bear" (*ursus thibetanus*), and the "snow bear" (*ursus isabellinus*).

The first mentioned is the one which has received most notice — both from naturalists and travellers. It is that species which by certain wiseacres of the closet school was for a long time regarded as a sloth (*bradypus*). In redeeming it from this character, other systematists were not content to leave it where it really belongs — in the genus *ursus* — but must, forsooth, create a new one for its special accommodation; and it now figures in zoölogical catalogues as a *prochilus* — the *prochilus labiatus!* We shall reject this absurd title, and call it by its real one — *ursus labiatus,* which, literally translated, would mean the "lipped bear" — not a very specific appellation neither. The name has been given in reference to a peculiar characteristic of the animal — that is, its power of protruding or extending the lips to seize its food — in which peculiarity it resembles the tapir, giraffe, and some other animals. Its trivial name of "sloth bear" is more expressive: for certainly its peculiar aspect — caused by the long shaggy masses of hair which cover its neck and body — gives it a very striking resemblance to the sloth. Its long crescent-shaped claws strengthen this resemblance. A less distinctive name is that by which it is known to the French naturalists, "ours de jongleurs," or "juggler's bear." Its grotesque appearance makes it a great favorite with the Indian mountebanks; but, as many other species are also trained to dancing and monkey-tricks, the name is not characteristic.

This bear is not quite so large as the *ursus arctos;* though individuals are sometimes met with approaching the bulk of the latter. The fur is longer and "shaggier" than in any other species — being upon the back of the neck full twelve inches in length. In this mass of long hair there is a curious line of separation running transversely across the back of the neck. The front division falls forward over the crown, so as to overhang the eyes — thus imparting to the physiognomy of the animal a heavy, stupid appearance. The other portion flaps back, forming a thick mane or hunch upon the shoulders. In old individuals the hair becomes greatly elongated; and hanging down almost to the ground on both flanks, and along the neck, imparts to the animal the strange appearance of being without legs!

The general color of the coat is black, with here and there a dash of brown over it. Upon the breast there is a white list of a triangular shape; and the muzzle is also a dirty yellowish white. There is no danger of mistaking this species for any other of the black Asiatic bears, or even any black bears. The long shaggy hair, hanging loosely, presents an appearance altogether different from the uniform brush-like surface, which characterizes the coats of *ursus malayanus, euryspilus, americanus, ornatus,* and *frugilegus.*

Perhaps the most peculiar characteristic of the sloth bear is the capability it possesses of protruding the lips, which it can do to a length of several inches from its jaws — shooting them out in the form of a tube, evidently designed for suction. This, together with the long extensile tongue — which is flat-shaped and square at the extremity — shows a peculiar design, answering to the habits of the animal. No doubt the extraordinary development of tongue is given to it for the same purpose as to the *edentata* of the ant-eating tribe — to enable it to "lick up" the *termites.*

Its great curved claws, which bear a very striking resemblance to those of the ant-eaters — especially the large *tamanoir* of South America — are used for the same purpose: that of breaking up the glutinous compost with which the termites construct their curious dwellings.

These insects constitute a portion of the sloth bear's "commissariat of subsistence;" but he will also eat fruits, and sweet succulent vegetables; and, it is scarce necessary to add, that he is "wild after" honey, and a regular robber of beehives.

Notwithstanding the comic *rôle,* which he is often taught to play in the hands of the jugglers, he not unfrequently enacts a little bit of tragedy. This occurs when in his wild or natural state. He is not disposed wantonly to make an attack upon human beings; and if left unmolested, he will go his way; but, when wounded or otherwise provoked, he can show fight to about the same degree as the black bear of America. The natives of India hold him in dread: but chiefly on account of the damage he occasions to their crops — especially to the plantations of sugar-cane.

We have stated that the sloth bear is not exclusively confined to the Himalayas. On the contrary, these mountains are only the northern limit of his range — which extends over the whole peninsula of Hindostan, and even beyond it, to the island of Ceylon. He is common in the Deccan, the country of the Mahrattas, Sylhet, and most probably throughout Transgangetic India. In the mountains that bound the province of Bengal to the east and west, and also along the foot-hills of the Himalayas of Nepaul on its north, the sloth bear is the most common representative of the Bruin family; but up into the higher ranges he does not extend his wanderings. His habitat proves that he affects a hot, rather than a cold climate — notwithstanding the great length of the fur upon his coat.

One peculiarity remains to be mentioned. Instead of hiding himself away in solitudes, remote from human habitations, he rather seeks the society of man: not that he is fond of the latter; but simply that he may avail himself of the results of human industry. For this pur-

pose he always seeks his haunt near to some settlement
— whence he may conveniently make his depredations
upon the crops. He is not, strictly speaking, a forest
animal. The low jungle is his abode ; and his lair is a
hole under some overhanging bank — either a natural
cavity, or one which has been hollowed out by some
burrowing animal.

Knowing that the sloth bear might be met with in any
part of the country, to the northward of Calcutta, our
hunters determined to keep a lookout for him while on
their way to the Himalayas — which mountains they intended ascending, either through the little state of Sikkim,
or the kingdom of Nepaul.

Their route from Calcutta to the hills lay a little to the
west of north ; and at many places on their journey they
not only heard of the sloth bear, but were witnesses of
the ravages which this destructive creature had committed
on the crops of the farmers.

There were sugar plantations, on which they saw tall
wooden towers raised in the middle of the field, and
carried to a considerable height above the surrounding
vegetation. On inquiring the purpose of these singular
structures, they were informed that they were intended
as watch-towers ; and that, during the season, when the
crops were approaching to ripeness, *videttes* were stationed
upon these towers, both by night and by day, to keep a
lookout for the bears, and frighten them off whenever
these plunderers made their appearance within the boundaries of the field.

Notwithstanding the many evidences of the sloth bear's
presence met with throughout the province of Bengal,

our hunters failed in falling in with this grotesque gentleman, till they were close up to the foot of the Himalaya
mountains, in that peculiar district known as the *Teräi*.
This is a belt of jungle and forest land — of an average
width of about twenty miles, and stretching along the
southern base of the Himalaya range throughout its
whole length, from Affghanistan to China. In all places
the **Teräi** is of so unhealthy a character, that it can
scarcely be said to be inhabited — its only human denizens being a few sparse tribes of native people (Mechs);
who, acclimated to its miasmatic atmosphere, have nothing to fear from it. Woe to the European who makes
any lengthened sojourn in the Teräi! He who does will
there find his grave.

For all its unhealthiness, it is the favorite haunt of
many of the largest quadrupeds : the elephant, the huge
Indian rhinoceros, the lion and tiger, the jungly ghau or
wild ox, the sambur stag, panthers, leopards, and chee-
tahs. The sloth bear roams through its thickets and
glades — where his favorite food, the white ants, abounds;
and it was upon reaching this district that our hunters
more particularly bent themselves to search for a specimen of this uncouth creature.

Fortunately they were not long till they found one —
else the climate of the **Teräi** would soon have so enfeebled them, that they might never have been able to climb
the stupendous mountains beyond. Almost upon entering within the confines of this deadly wilderness, they
encountered the sloth bear; and although the interview
was purely accidental, it ended in Bruin being deprived
of his life and his long-haired robe.

15   v

The sloth bear did not submit tamely to this double robbery, for he was himself the assailant — having been the first to cry " Stand and deliver !" Nor was his conquest accomplished without a perilous struggle — that came very near reducing the number of our heroes from odd to even. But we shall give the account of the affair, as we find it detailed in the journal of Alexis.

CHAPTER LX.

BRUIN TAKEN BY THE TONGUE.

THE travellers had halted for lunch, and tied their horses to the trees. While Pouchskin was spreading out the comestibles, and Alexis engaged in noting down in his journal the events of the day, Ivan — attracted by a beautiful bird — had taken up his fowling-piece, and followed the bird through the jungle — in hopes of getting a shot at it. We go along with Ivan, for it was he who started the "mountebank" bear, that came near mounting him on the moment of their meeting it.

Ivan was walking cautiously along a bank, that rose to about the height of his head; but which in places was undermined, as if by the action of running water — though there was no water to be seen. The ground, however, upon which he trod was covered with pebbles and coarse gravel — showing that at some period water must have flowed over it; and, indeed, it was evidently the bed of a stream that had been full during the rainy season, but was now completely dried up.

Ivan was not thinking of this; but of the beautiful bird which was flitting about among the trees — still keeping out of the range of his gun. He was in a bent attitude, crouching along under the bank — which he was using as a cover, to enable him to approach the tantalizing game.

All at once, a singular noise fell upon his ear. It was a sort of monotonous purring, like that made by a spinning-machine, or a very large tom-cat ; and like the latter, it was prolonged and continuous. The sound was not exactly pleasant to Ivan's ear, for it denoted the proximity of some animal ; and, although it was not loud, there was something about the tone that told him the animal giving utterance to it was a creature to be feared. In fact, it fell upon Ivan's ear in the character of a warning ; and caused **him to** desist from his pursuit of the bird, come suddenly to a stand, and listen with great attention.

For some moments he was unable to make out whence the sound proceeded. It seemed to fill the space all around him — as if it came out of the air itself — for the purring sound kept the atmosphere constantly vibrating ; and, as there was no definite concussion, it was all the more difficult to trace it to its source.

The thought that had entered into Ivan's mind was that it might be the purring of a tiger he heard ; and yet it seemed scarcely so harsh as that — for he knew the peculiar rattle which frequently proceeds from the thorax of the royal Bengalese cat.

He quickly reflected, however, that whether it was tiger or not, it would neither be safe for him to raise an alarm, nor start to rush back to the bivouac — though this was not twenty yards from the spot. By making an attempt to retreat, he might draw the animal after him, or stumble upon it — not knowing its direction. It was to ascertain its whereabouts that he had stopped and stood listening. That once known, he might keep his place, or take to flight — as circumstances should dictate.

Nearly a minute remained he in this irresolute attitude — looking around on every side, and over the bank into the contiguous jungle; but he could see no living thing of any kind — for even the bird had long since taken its departure from the place. Still the purring continued; and once or twice the sound increased in volume — till it almost assumed the character of a " growl."

All at once, however, it came to an end; and was succeeded by a quick, sharp " sniff," several times repeated. This was a more definite sound; and guided Ivan's eyes in a direction in which he had not before thought of looking. He had hitherto been reconnoitring around him and *over* the bank. He had not thought of looking *under* it.

In this direction were his eyes now turned; and stooping his body, he peered into the dark subterraneous excavation which the water had caused in the alluvial earth. There, to his surprise, he beheld the author of the baritone performance that had been puzzling him.

At first he saw only a countenance of a dirty-whitish color, with a pair of ugly, glancing eyes; but, in looking more attentively, this countenance was seen to protrude out of an immense surrounding of black shaggy hair, which could be the covering of no other animal than a bear — and a sloth bear at that?

On making this discovery, Ivan did not know whether to be merry or sad. He would have been glad enough, had he seen the bear at a distance; but, situated as he was — with the great brute near enough to reach him at

a single spring, — in fact, almost between his legs, — he
had little cause to congratulate himself upon the "find."
Nor did he. On the contrary, he was seized with a
quick perception of danger, and only thought of making
his escape. He would have turned upon the instant and
fled ; but it occurred to him, that by doing so he would
draw the bear after him, and he knew that, notwithstanding the uncouth shuffle which a bear makes in running,
— and the sloth bear is the greatest " shuffler " of the
family, — he can still go too fast for a man. Should he
turn his face, the bear might spring upon his back, and
thus have him at his mercy.

Instead of facing away, therefore, Ivan kept his front
to the bank : and with his eyes fixed upon the animal,
commenced gliding backwards slowly but silently. At
the same time he had cautiously raised his gun to the
level — with no intention, however, of firing, but merely
to be ready in case the bear should become the assailant. Otherwise, Ivan was perfectly agreeable to making it a "draw" between them.

Bruin, however, had no idea of thus giving up the
game; for the fierce growl which just at that moment
escaped him, signified anything but assent. On the contrary, it was the prelude to the play; and declared his
intention of beginning it. Almost simultaneous with the
growl, he was seen starting to his feet ; and before Ivan
could pull trigger, or even raise his gun to a proper elevation, a huge mass of black, shaggy hair, like a bundle
of sooty rags, came whisking through the air directly
towards him. Men talk of the sudden spring of the tiger,
and the quick, rushing charge of the lion; but strange as

it may seem, neither one nor other of these animals can charge forward on their intended victim with more celerity than a bear — clumsy and uncouth as Bruin may appear. His capacity of raising himself erect gives him this advantage; and from his great plantigrade posterior paws, combined with his powerful muscular legs, he can pitch forward with a velocity surprising as it is unexpected. This the regular bear-hunter well knows; and the knowledge renders him cautious about coming too close to a *couchant* bear. Ivan himself knew it; and it was for this very reason he was endeavoring to widen the distance between himself and Bruin, before he should turn to run.

Unfortunately he had not succeeded in gaining sufficient ground. He was still within charging distance of the animal as it rose to its feet; but another step backward as the bear launched forth, carried him clear of the spring; and Bruin leaped short. In another instant, however, he erected himself, and again sprang forward; but this time the impetus given to his body was not so great; and although he succeeded in closing with the young hunter, the latter was enabled to keep his feet and grapple with him in an erect attitude. Had he fallen to the ground, the bear would have made short work with him.

Ivan had dropped his gun; for, not having time to raise it or take aim, the weapon was of no use. His hands were therefore free; and as the bear pitched up against him, he stretched out his arms, grasped the long hair that hung over the frontlet of the animal, and with all his might held back the monster's head with his threatening jaws.

The bear had thrown both his paws around the body of the young hunter; but a broad thick belt which the latter chanced to have on, protected his skin from the animal's claws. So long as he could hold back that open mouth, with its double rows of white, sharp teeth, he had not so much to fear; but his strength could not last long against such a powerful wrestler. His only hope was that the cries which he was raising would bring the others to his assistance; and of this he had no doubt: as he already heard both Pouchskin and Alexis hurrying up towards the spot.

It was a perilous moment. The extended jaws of the bear were within twelve inches of the young hunter's face; he could feel the hot breath steaming against his cheeks, and the long extensile tongue almost touched his forehead, vibrating about in rapid sweeps, as if the animal by that means hoped to bring his head within reach!

The struggle was not protracted. It lasted till Alexis and Pouchskin came upon the ground; but not six seconds longer. The first thing that Pouchskin did was to grasp the protruding tongue of the bear in his left hand — making a half curl of it round his fingers — while with his right he plunged his long knife right between the ribs of the animal. Alexis, on the other side, dealt a blow in similar fashion; and, before either of them could draw his blade out of its hair-covered sheath, the huge mountebank relaxed his hold, and rolled over among the pebbles. There, after a few grotesque contortions his limbs lay extended and motionless, making it evident beyond a doubt, that *his* "dancing days were over."

CHAPTER LXI.

AN EXTRA SKIN.

OUR hunters did not remain at their bivouac longer than was absolutely necessary to swallow a hasty meal. They had been warned of the dangerous climate of the *Terái*, and hurrying on through it, reached the more elevated hill region before night.

Journeying on, they entered the kingdom of Nepaul, among whose hills they expected to find the Thibet bear (*ursus thibetanus*). This animal has been usually regarded as a mere variety of the *ursus arctos ;* but without the slightest reason. It is an animal of more gentle habits, and exclusively a vegetarian in its diet : in color it is black, but having a white mark on its breast shaped like a **Y, the** branches of the letter coming up in front of its shoulders, while the limb extends between the fore legs and half-way along the belly. The claws of the animal are small and weak ; and its profile forms almost a straight line, thus essentially differing from the *ursus arctos.* It is also a much smaller animal — rarely attaining to more than half the size of the latter species, and scarce bigger than the *ursus malayanus*, to which it bears a far greater resemblance. It is found in the mountains of Sylhet, and throughout that portion of the

15 *

Himalayas enclosed within the great bend of the Brah-
mapootra, in Thibet, whence it derives its specific appellation. It is equally an inhabitant of the hill-country of
Nepaul; and there our hunters proceeded in search of
their specimen. By the help of a "Ghoorka" guide,
which they had hired, they were not long in finding one;
but as there was no curious or particular incident connected with its capture, the journal of Alexis is silent
upon the affair: it is only recorded that the animal was
started from a thicket of *rhododendron* bushes, and shot
down while endeavoring to make its escape.

Having settled their business with the Thibet bear,
our hunters might have also procured another species
within the territory of Nepaul — that is, the brown, or
Isabella bear (*ursus isabellinus*). This they could have
found by ascending to the higher ranges of the great
snowy mountains that overlook Nepaul; but as they
knew they should also encounter this species near the
sources of the Ganges, and as they were desirous of
visiting that remarkable locality, they continued on westward through Nepaul and Delhi, arriving at the health
station of Mussoorie, in the beautiful valley of the Dehra
Doon.

After resting here for some days, they proceeded to
ascend the mountains, the lower and middle zone of
which they found covered with forests of magnificent
oaks, of several distinct species.

In these oak forests, greatly to the surprise of Alexis,
they heard of the existence of a large, black bear, altogether different from the *ursus thibetanus*, and equally
so from the *ursus isabellinus* — a distinct species, in fact,

which, though well known to Anglo-Indian hunters, appears to have escaped the attention of naturalists.

They ascertained, moreover, that he was far from being a scarce animal, or an insignificant member of the Bruin family ; in point of size, formidable strength, and ferocity of disposition, being only inferior to *ursus ferox* and *maritimus*, and in all these qualities quite a match for the *ursus arctos*. Of his fierce nature, and the capability to do mischief, our travellers had evidence in almost every village through which they passed. Numerous instances were brought before their notice of men who had been scratched and torn by these black bears, and some most fearfully mutilated. They saw men with their whole skin stripped from their skulls and faces ; their features presenting a most hideous aspect.

This singular habit of inflicting punishment on their human enemy appears to be common to the whole bear tribe — I mean, the habit of scalping their victims, and endeavoring to disfigure the face. Not only do both the black and brown bears of the Himalayas follow this habit, but also the *ursus arctos*, the grisly, and the white. They always aim at the head, but more especially the face ; and with a single " rake " of their spread claws, usually strip off both skin and flesh.

Having accomplished this, a bear will often desist from further ill-treatment of his victims ; and if the latter will but lie still and feign dead, the monster will give up mauling him, and shamble off from the ground, apparently satisfied with having taken the scalp.

This savage habit on the part of the bears our young

hunters had long since noted; and that the black bear of
the Himalayas followed the fashion of his kindred, they
had now ample evidence.

In his other habits — which they learnt from the Shikkaries, or village hunters — this bear strongly resembles
the *ursus arctos* of Northern Europe. On ordinary occasions his food consists of fruits, roots, and insects of every
kind he can catch — even scorpions and beetles — and
where the primeval forest does not afford him full rations,
he will enter the cultivated grounds and make havoc
among the crops. Strange enough, he does not meddle
with the wheat; though he will ravage the fields of
buckwheat and barley! At night he enters the gardens
contiguous to the houses, and plunders them of all kinds
of fruits and vegetables. He even approaches still
nearer — abstracting their honey from the tame bees —
the hives of which, according to a curious custom of the
hill people, are set in little indentations in the walls of
their dwelling-houses.

The black bear occasionally cools his chops by munch-
ing melons and cucumbers; but he is particularly fond
of a dessert of apricots — which is the most common
fruit cultivated throughout the middle ranges of the **Him-**
alayas. The bear enters the apricot orchard at night;
and climbing the trees, will make more havoc in a single
visit than a score of school-boys. In all the orchards,
elevated crows'-nests or sentry boxes are set up, specially
intended for watching the bears; and at this season many
of them are killed in the act of robbing.

The Himalayan black bear will eat flesh — either fresh
or putrid — and when once he has got into this habit he

never forsakes it, but remains a carnivorous creature for the rest of his life. He will attack the goats and sheep on the mountain pastures; and will even make inroads to the village enclosures, and destroy the animals in their very sheds! When a **flock of** sheep falls in his way, unless he is driven off by the shepherds, he does not content himself by killing only one, but sometimes converts a score of them into mutton.

Those bears, however, that exhibit an extreme carnivorous propensity, are certain to bring about their own destruction: as the attention of the villagers being drawn upon them, snares and baited traps are set everywhere, and they are also followed by the Shikkaries armed with their matchlock guns.

These bears often attain to an immense size — in this respect nearly equalling the *ursus arctos,* of which they cannot, however, be supposed to be a variety. Eight feet is the usual length of a full-grown specimen; and, when in a good condition, it requires a whole crowd of men to raise the carcass of one of them from the ground.

Autumn is their season of greatest fatness; and especially when the acorns are getting ripe, but previous to their falling from the tree. Then the black bears are met with in the greatest numbers, coming from all parts into the oak forests, and climbing the trees to procure their favorite food. They do not nibble off the acorns one by one; but first break the branches which are loaded, and carry them all into one place — generally into some fork — where, seated like squirrels, on their great hams, they can discuss the meal at their leisure.

In passing through these oak forests, large piles of branches may be seen thus collected together on the tops of the trees — resembling the nests of rooks or magpies — which have been brought together by the bears for the purpose above stated.

When the forest lies in a district where these bears are much hunted, they usually retire by day; and conceal **themselves** in their hiding-places in the thickets; but even in such forests the animals may be seen prowling about before sunset, and long after daylight in the morning.

In the higher hills and forests of the *khurso* oak, remote from the villages, they do not even take the precaution to hide themselves, but remain all day "acorn-gathering" among the trees. It is at this season that they can be hunted with most success: since the hunter is under no necessity of tracking them, but can find his great game by simply walking quietly through the woods, and keeping a lookout overhead, just as if he were searching for squirrels.

It chanced to be the month of October when our hunters arrived at this part of the Himalayas; and having reached the region of the larger oak forests, they commenced their search accordingly. They were extremely desirous of success; knowing how much their father would be gratified at obtaining the skin of this black bear, which being an undescribed variety, might be considered an "extra" one.

CHAPTER LXII.

AN UNHAPPY HORSE.

Our young hunters commenced their search in a forest of *khurso* oaks, which, interspersed with cedars and other trees, covered a high, round-topped ridge, that rose above the little village where they had made their head-quarters.

On reaching the flat summit of the ridge, they found they could manage better without their horses : as seated in the saddle they could not so well reconnoitre the tops of the trees, where they expected to see their game. They dismounted, therefore, and leaving their animals tied to the branches of a large spreading cedar-tree (the *deodor*), they proceeded onward on foot.

On this day the luck seemed to be against them ; for although they met with plenty of " signs " — where the bears had broken the branches of the oaks — and also saw numbers of freshly-made " rooks'-nests," they could not get their eyes upon Bruin himself, who had left these tokens of his presence. It might be that this forest was frequently hunted by the native Shikkaries ; and that would account for the absence of the bears during the daytime. They had gone, no doubt, to their hiding-places.

This was the conclusion at which our hunters arrived — after tramping about until they were tired; and not having met with a single bear.

It was now the hour of noon; and, as they had been told that the evening would be the likelier time **to** find Bruin upon the prowl, they resolved returning to where they had left their horses, and remaining there until evening should arrive. They had grown hungry; and, having walked many miles, were pretty well done up. A bit of dinner, and a few hours' rest under the great cedar, would recruit their strength; and enable them to take the field again before sunset with a better prospect of success.

Following their back-track through the forest, therefore, they proceeded towards the place where they had left their horses.

Before coming in sight of these animals, they were admonished of their proximity by hearing them neighing at short intervals; but, what surprised them still more, they heard a constant pounding — as if the horses were striking the **ground** repeatedly and continuously with their hoofs!

Arriving within view of them, their astonishment was not-diminished, on perceiving that the three horses were rearing and dancing over the ground, as if endeavoring to break loose from their fastenings! Each had been tied to a separate branch of the tree — their bridles being simply noosed over the twigs at the extremities of the branches; and allowing them to play to the full length of the rein. Consequently, the three horses were many yards apart from **each** other; **but all were** equally in mo-

tion — all neighing and pitching about, as if something
had set them mad!

Could it be horse-flies? thought the hunters. They
knew there was a species of horse-fly in the Himalayas
— greatly dreaded by all animals, and even by man himself. They knew this: for they had already suffered
from its persecuting bite. But this was in the lower
valleys; and it was not likely it should be found at the
elevation of this *khurso* forest — quite 10,000 feet above
sea level.

Perhaps bees? There might be a nest of wild bees
somewhere near — why not in the cedar itself — and if
so, the horses might be attacked by them? That would
account for the capers they were cutting!

They had almost settled it in their mind that this
was the true explanation; when an object came before
their eyes that gave a very different solution to the mystery.

One of the horses appeared more frightened than the
other two — at least he was squealing and curvetting in
a much more violent manner. As he danced around,
his eyes appeared to be directed upwards — the great
eyeballs sparkling, and protruded as if about to start
from their sockets. This guided the glances of the hunters; and, looking among the branches of the cedar, they
now perceived a large black mass, of an oblong shape —
extended along one of the lower limbs, and just over the
spot where the horse was tied.

They had hardly time to make out the shape of this
dark object, and become convinced that it was the body
of a bear, when **the huge** creature was seen to launch

itself down from the limb; and then drop like a cat, all-fours, upon the back of the horse!

The latter uttered a scream of affright; and as if terror had added to his strength, he now succeeded in breaking the branch — around which the rein was looped — and bounded off through the forest, the bear still squatted upon his **back**!

The trees that stood around were nearly all of slender growth; but, as their stems grew thickly together, the **horse, with** his strange rider, could make but slow way among them; and every now and then the former, half blind with affright, dashed his sides against the trunks, causing them to crackle and shiver at each concussion.

All at once the horse was seen coming to a halt, as if brought up by the power of a Mameluke bit! The spectators saw this with wondering eyes — unable for the moment to explain it. As they were very near the spot where the halt had been made, they soon perceived the nature of the interruption. The bear had thrown one of his great fore-arms around a tree; while, with the other, he still clutched the horse, holding him fast! The design of Bruin was perfectly clear: he had seized the tree in order to bring the steed to a stand!

In this for a time he was successful. With one arm he was enabled to retain the tree in his powerful hug; while with the other he held the horse — his huge paw, with its retentive claws, being firmly fixed under the pommel of the saddle.

A singular struggle now ensued, which lasted for some seconds of time; the horse making the most energetic efforts to escape; while the bear was equally eager **in** endeavoring to retain **him**.

Lucky was it for the steed that his master was not more particular about the girth of his saddle, and that either the strap or buckle was a bad one. Whichever of the two it was, one of them gave way; and the horse, thus freed, was not slow to profit by the fortunate accident. Uttering a neigh of joy, he sprang onward — leaving both bear and saddle behind him.

So far as the horse was concerned, his danger was over. Not so with the bear, whose troubles were just now to begin. While holding the horse in his muscular arm — and clutching the pine with the other — the tree had got bent until its top almost touched the saddle. When the girth broke, therefore, the elastic sapling sprang back like a piece of whalebone; and with such an impetus as not only to shake Bruin from his hold, but to pitch him several yards to the opposite side — where he lay stunned, or at all events so astonished, as, for a moment, to appear as if he had taken leave of his life!

This moment of the bear's embarrassment was not lost upon the hunters, who ran rapidly up — till within ten paces of the prostrate animal — and discharging their guns into his body, prevented him from ever again getting to his feet. His hide was the only part of him that afterwards attained the erect attitude; and that was when it was mounted in the museum of the palace Grodonoff.

CHAPTER LXIII.

THE SNOW BEAR.

Higher up the Himalayas dwells the " snow bear." This species has received from naturalists the very fanciful appellation of the " Isabella bear " (*ursus isabellinus*) — a title suggested by its color being that known as " Isabella color," — the type of which was the very dirty gown worn by Queen Isabella at the siege of Grenada. It is doubtful whether any living man could exactly tell what is an Isabella color ; and the use of such a phrase in describing the hue of an animal's skin is altogether indefinite and, to say the least, absurd.

The " Isabella bears," moreover, are not always of the so-called Isabella color. On the contrary, there are some of dark-brown, some of a hoary-brown, and others nearly white ; and to Himalayan hunters they are known by the various appellations of brown, red, yellow, white, gray, silver, and snow, showing the numerous varieties of color met with in the species. Some of these varieties are to be attributed to the different seasons of the year and the age of the animal.

Of all these designations, that of " snow bear " appears the most characteristic, since it avoids the risk of a confusion of names — the other titles being equally bestowed upon certain varieties of the *ursus americanus* and *ursus ferox*. It is also appropriate to the Himalayan animal: since his favorite haunt is along the line of perpetual snow; or in the grassy treeless tracts that intervene between the snow-line and the forest-covered declivities — to which they descend only at particular times of the year.

In identifying this species, but little reliance can be placed on color. In spring their fur is long and shaggy —of various shades of yellowish-brown, sometimes reddish-brown, and not unfrequently of a gray or silvery hue. In summer this long, yellowish fur falls off; and is replaced by a shorter and darker coat, which gradually grows longer and lighter as the winter approaches. The females are a shade lighter-colored than the males; and the cubs have a broad circle of white around the neck, which gradually disappears as they grow to their full size.

The snow bear *hybernates*, hiding himself away in a cave; and he is only seen abroad when the spring sun begins to melt the snow upon the grass-covered tracts near the borders of the forest. On these he may be found throughout the summer — feeding upon grass and roots, with such reptiles and insects as come in his way. In the autumn he enters the forests in search of berries and nuts, and at this season — like his congener, the black bear — he even extends his depredations to the cultivated grounds and gardens of the villagers, in search of fruit and grain, buckwheat being a favorite food with him.

Though naturally a vegetarian in his diet, he will eat flesh-meat upon occasions; and frequently makes havoc among the flocks of sheep and goats, that in summer are

taken up to pasture on the grassy tracts above mentioned.
While thus engaged, he does not regard the presence of
man ; but will attack the shepherds who may attempt to
drive him off.

Among the many strange items that compose the larder
of the snow bear, grubs and scorpions have a prominent
place. He spends much of his time in searching for
these — scratching them out of their holes, and turning
over stones to get at them. Great boulders of rock, that
a man could not move, he will roll over with his muscular arms ; and large tracts of ground may be seen with
the stones thus displaced.

It was while engaged in this curious occupation, that
our hunters came upon one of the snow bears ; which
they succeeded in killing. He was not the first they
had encountered : they had started several, and wounded
two ; but both had got off from them. This one, however, fell to their " bag," and in rather an unexpected
fashion.

They were working their toilsome way up a narrow
ravine — which, although the season was autumn, was
still filled with snow, that lay in the bottom of the gorge
to a great depth. It was snow that had lain all the year ;
and although not frozen, the surface was firm and stiff ;
and it was with difficulty they could get support for their
feet on it. Here and there they were compelled to stop
and cut steps in the snow — as the surface-sloped upward at an angle of full 50°, and, in fact, they were
rather climbing than walking. Their object, in undertaking this toilsome ascent, was simply because they had
seen a bear going up the same way but a few minutes

before ; and the scratches of his claws were visible on the
snow just before their faces.

Making as little noise as possible, they kept onward ;
and at length reached the head of the gorge. On peep-
ing cautiously over, they saw a little table-like tract of
level ground, several acres in extent. It was quite clear
of snow; and covered with green herbage. A number
of large boulder stones lay scattered over it — which had
evidently rolled down from the mountain-side that rose
still higher above the table.

But the sight that most gratified them was the bear
himself — no doubt, the same they had seen going up
the ravine. They now discovered him upon the level
ground, not twenty yards from the spot where they stood.
In a strange attitude they saw him — grasping between
his fore paws a huge boulder stone, almost as large as
his own body, and evidently in the act of rolling it out
of its bed !

They were the less astonished at what they saw : for,
being already acquainted with this singular habit of the
snow bear, they knew what he was about. They did not
stay, therefore, to watch his herculean labors ; but all
three, levelling their guns, pulled trigger simultaneously.
The bullets — some of them, at least — evidently struck
the bear ; but, although he dropped the great boulder —
which at once fell back into its place — he did not himself drop. On the contrary, he turned suddenly round ;
and, giving utterance to a savage growl, rushed direct
towards the hunters.

The latter, not having time to reload, had no choice
but to run for it. There was no other way of escape

open to them, except by the gorge up which they had
come ; as, to attempt ascending to the level ground would
have brought them face to face with the bear. They
turned, therefore ; and commenced retreating down the
ravine.

But now came the difficulty. They had not made
three strides, before perceiving that they could not keep
their feet upon the hard sloping surface of the snow.
They had no time to cut fresh steps, nor pick out their
old ones : as by doing either they would go too slowly,
while the bear could scramble down the snow as rapidly
as on bare ground. There was no alternative, therefore,
but to fling themselves on their posteriors, and slide down
the slope.

Quick as came the thought, all three of them dropped
down upon their hams ; and using their guns to prevent
them from going with too great velocity, they shot downward to the bottom of the ravine.

On reaching the lower end of the slope, and regaining
their feet, they turned and looked back up the gorge.
The bear had arrived at the upper end ; and was standing with his fore feet projected over the edge, and resting
upon the snow. He appeared to be undecided, as to
whether he should come down after them, or give up the
pursuit. He was within easy range of a bullet ; and
they bethought them of reloading and giving him a fresh
volley ; when, to their chagrin, they discovered that the
barrels of their guns were filled with snow — which had
got into them during the descent.

While lamenting this unfortunate accident — in the
full belief that they would now lose the bear — they saw

the animal make a strange movement. It was forward, and towards them — as if he had made up his mind to charge down the slope; but they soon perceived that this could not be his intention: for as he came gliding on, sometimes his head, and sometimes his stern was foremost; and it was evident that instead of the movement being a voluntary act on his part, it was quite the contrary. The fact was, that the bullets which they had fired into him had drawn the life's blood out of his veins; and having stood too long on the sloping edge of the snow, he had fallen through feebleness; and was now tumbling down the ravine, without strength enough to stay his descent.

In another instant he lay stretched almost at the feet of the hunters; for the impetus imparted to his huge carcass in the descent, had brought with it such a "whack" against a large rock, as to deprive him of whatever either of blood or breath there had been left in his body.

The hunters, however, made sure of this, by drawing their long knives, and making an additional vent or two between his ribs — thus securing themselves against all risk of his resuscitation.

They had now finished with the Himalayan bears of known and unknown kinds; but Alexis learnt enough from hunters, whom they had encountered during their sojourn in these mountains, to convince him that great confusion exists among naturalists as to the different species and varieties that inhabit the Himalayan range. Of the "snow bear" itself, a variety exists in the mountains of Cashmere; which, as far as Alexis could learn,

16

was very different from the kind they **had** killed. The Cashmerian variety is of a deep reddish-brown color, much longer in the muzzle than the "snow bear," and also a more dangerous antagonist to man — being a brute of eminently carnivorous propensity and savage disposition.

"It is quite probable," remarks Alexis, in his **journal,** "that instead of three kinds of bears inhabiting the Himalayan range, twice that number of "species" — or at all events, of permanent varieties — may be found within the extensive area covered by these stupendous mountains."

CHAPTER LXIV.

THE LAST CHASE.

OUR travellers descended once more to the plains of Hindostan, and crossed the peninsula by *dâk* to Bombay. From Bombay they sailed through the Indian Ocean, and up the Persian Gulf to the port of Bussora, on the Euphrates. Ascending the Tigris branch of this Asiatic river, they reached the famed city of Bagdad. They were now *en route* for the haunts of the Syrian bear among the snowy summits of Mount Lebanon. With a Turkish caravan, therefore, they started from Bagdad; and after much toil and many hardships, arrived in the city of Damascus — the scene of so many troubles and massacres caused by the fanaticism of a false religion.

With these questions our travellers had nothing to do; nor did they stay any length of time within the walls of the unhappy city. Soon after their arrival in the place, they obtained all the information they required of the where-abouts of the Syrian bear; and their steps were now direct-ed towards the snowy summits of Libanus — better known to Christians by its Scriptural name of Mount Lebanon.

In these mountains the Syrian bear (*ursus syriacus*) is found; and it is only a few years since the animal was discovered there. Every naturalist had doubted the

existence of bears in any part of Syria — as they now
deny that there are any in Africa. Those who acknowl-
edge it, are inclined to regard the Syrian bear as a mere
variety of the *ursus arctos ;* but this theory is altogether
incorrect. In shape, color, and many of his habits, the
Syrian bear differs essentially from his brown congener ;
and his dwelling-place — instead of being in forest-
covered tracts — is more generally in open ground or
among rocks. In fact, his range upon the Syrian mountains is very similar to that of the "snow bear" on the
Himalayas — near the line of perpetual snow.

The color of the *ursus syriacus* is a light ash or
fulvous brown, often with a hoary or silvery tinge —
but the color varies at times to lighter and deeper shades.
The hair lies close against the skin — in this respect
differing from most of the species, in which the fur stands
erect or perpendicular to the outlines of the body.
This gives the Syrian bear the appearance of being a
thinner and smaller animal than many bears of upright
fur that are no bigger than he.

By one characteristic mark he may be easily identified ; and that is, by his having an erect ridge of fur
running from his neck along the spine of his back, and
looking not unlike the mane of a donkey. But, indeed,
the Syrian bear may be easily distinguished from any
other member of this family ; and to regard him as a
mere variety of the *ursus arctos,* is only going back to
the old system that considers all the bears as one and the
same species.

The Syrian bear does not inhabit the whole range of
the mountains that pass under the general name of Leb-

anon. **Only** on the loftier summits is he found — particularly on that known as Mount Makmel. This summit is covered with snow ; and it is under the snow-line he usually makes his haunt. Sometimes, however, he descends to a lower elevation ; and in the village gardens — just as does the snow bear in **the** Himalayas — he make sad havoc among fruits and vegetables. He will also kill sheep, goats, and even larger animals that come in his way ; and when provoked will attack the hunter without fear. **He is** most dreaded in the night : for it is during **the** darkness he generally makes his plundering expeditions. Both shepherds and hunters have been killed by him — proving that he still retains the savage character given to him in the Scriptures ; where several **of his** kind — she-bears they **were** — are represented as having torn " **forty and two of the mockers of** Elisha."

He appears to have been equally characterized by a ferocity of disposition **in the** crusading ages — since it is related that the great leader Godfrey slew one of these **bears, whom he** found assaulting a poor wood-cutter of **Antioch ; and the affair was** considered a feat of great prowess by those eccentric champions of the Cross.

That the Syrian bear is still as ferocious and savage as he ever could have been, our hunters proved by their own experience : for although they did not get into the power of one, they would certainly have done so — some one of them at least — had they not been **fortunate** enough **to** kill the bear before he could lay his claws upon them. But we shall briefly describe the adventure ; which was the last our hunters were engaged in — at least, the last we find recorded in the journal of Alexis.

Bischerre, a little mountain village, situated near the snow-line on Mount Makmel, had become their temporary head-quarters. Its neighborhood was celebrated for the great number of bears that frequent it. These animals descending from the higher ridges surrounding it, fre-frequently enter the gardens of the villagers, and rob them of their vegetables and chick-peas (*cicer arietinus*) — the latter being a favorite food of the Syrian bear.

From Bischerre the hunters extended their excursions on foot : since the nature of the ground would not ad-mit of their using horses ; and they had succeeded in getting several good " bear-chases," and in killing a brace of these animals. Both, however, were very young ones — cubs, in fact — and their skins would not do. A better specimen must be procured.

This came into their hands in the following man-ner : —

They had succeeded in tracing a bear up into a rocky ravine — the entrance into which was not over ten or twelve feet in width. The ravine itself was a steep descent leading up to the mountains ; and its bottom, or bed, was covered with a conglomeration of large rounded boulders, that looked as if they had been rolled into this shape by water. They resembled the round stones some-times seen in rivers ; and no doubt there was a torrent there at times ; but just then the channel was dry, and not a drop of water appeared anywhere. There was no snow either ; as the place was below the line of snow ; and they had only traced the bear into it on information given them by some shepherds, who had seen the animal recently enter it.

Relying upon this information, they kept up the defile, making their way with difficulty over the loose pebbles. They had a hope that the bear was still somewhere within the gorge ; and that they might find him in some crevice or cave. On each side rose high cliffs that almost met overhead ; and our hunters, as they scrambled up the steep, examined these cliffs carefully — expecting to perceive the mouth of a cavern. The place was likely enough, for at every few yards they saw crevices and deep cavities ; but in none of them could they find any traces of Bruin.

They had got about half way through the ravine — and were still scrambling upward — when a loud sniff drew their attention ; and, looking in the direction whence it appeared to have proceeded, there, sure enough, was the identical animal they were after — Master Bruin himself. They saw only his snout ; which was projected out from the face of the cliff, about twenty feet above the bed of the ravine. His whole head was shortly after poked forth ; and seen *en profile* from below, it looked as if there was a bear's head glued against the flat surface of the rock, just as stags' heads are seen ornamenting the halls of grand country mansions. Our hunters, however, knew there must be a cave behind — in which was the body of the bear, though it was concealed from their eyes.

The bear, after glancing at the intruders who had disturbed him, drew back his head so suddenly that not a shot could be fired in time. The hunters, in order to get into a better position, hurried **past under the** cave; and took stand several paces above it — where they were able to command a better view of the entrance.

They were now on a level with the hole out of which the head had shown itself; and without speaking a word, only in whispers, they waited for the reappearance of the snout.

It was not long before they had the satisfaction of seeing it. Whether from curiosity to know if they were gone — or with the design of sallying forth in pursuit of them — the bear once more protruded his muzzle from the hole. Fearing that he might **draw it** back again, and not **give** them another chance, all three fired, and in such haste that two of them quite missed the object. **Only** the bullet of Alexis had been properly aimed; and this was seen striking the bear right in the teeth — several of which were shot clean out of his jaws.

As the smoke cleared out of their eyes, the great yellow body of the bear was observed out upon the little ledge that projected in front of the cave; and uttering loud screams — expressive both of rage and pain — the angry animal bounded down among the boulders. Instead of making down the ravine — as our hunters expected — he turned upwards, and rushed directly towards them.

Again there was no alternative but flight; and up the steep gorge they must go. To make downward would be to run right **upon the claws** of the infuriated animal; and upward was the only way left open to them.

All three started and ran **as** fast as they were able; and for a while were in hopes of distancing their pursuer. But further up, the slope grew steeper; and the loose stones became more difficult to clamber over. Their breath, too, was by this time quite gone; and all three were panting like "winded" horses.

It was impossible for them to go a step farther.

In despair, they halted; and turned to face the pursuer — all of them at the same instant drawing their knives; and bracing their bodies for the expected struggle. The bear, still growling and screaming, came on — making way over the stones much faster than they had done. He would have been certain of overtaking them, had they continued their race: for he was scarce six paces behind them when they had stopped.

No doubt it would have been a dangerous conflict, had it come off; and, indeed, breathless as they were, they could never have sustained the attack. Of course, they had no time to reload their guns, and did not think of such a thing. Their determination was to defend themselves with their knives; and perhaps they might have succeeded in doing so, had there been an occasion. But there was not.

Before the bear could get up to them, a better idea had flashed across the brain of Pouchskin; which he lost not a moment in carrying into execution. Stooping suddenly, and flinging his knife out of his hands, he laid hold of a large boulder — big enough to weigh at least half a hundred — and raising this to the height of his shoulder, he hurled it down upon the bear.

The huge stone struck the animal right upon the breast; and what with the force by which it had been launched from Pouchskin's powerful arm, and the im-

petus it had gained in its descent, it acted on Bruin like
a thunderbolt — not only knocking him over on his back,
but carrying his body along with it full ten paces down
the gorge.

When the hunters at length reloaded their guns, and
went down to where Bruin lay among the rocks, they
found him lying doubled up as dead as mutton.

Having stripped him of his fulvous skin, they returned
to Bischerre; and next day packing up their *impedimenta*, they crossed through the passes of Mount Libanus,
and proceeded onward to the shores of the Mediterranean Sea.

Home was now the word; and right pleasant was the
sound of it in their ears. The grand bear-hunt was
ended. They had accomplished the task imposed upon
them — having kept every condition of their covenant.

Of course they expected a grand welcome upon their
return; and in this expectation they were not disappointed; for many days and nights after the baronial
halls of the palace Grodonoff, echoed the sounds of
mirth and revelry.

In the museum our young hunters met their old ac-
quaintances, from all parts of the world. They encountered them standing in different attitudes — all mounted
in the most approved fashion. The Syrian bear was
the only one not among them: as they had themselves
brought his skin — all the others having been sent home
by "Parcels Delivery." In a few days, however, the
ursus syriacus was set upon his legs; and the collection
was complete.

The news of the "Grand Bear-Hunt," with its curious

conditions, soon got abroad; and travelled all round the social circle of St. Petersburgh. Figuratively speaking, our young hunters were transformed into animals themselves — they became "lions," — and remained so for that season; but even at this hour in the *salons* of the great Russian capital, you may often hear introduced, as a favorite topic of conversation —

"THE BARON AND HIS BEARS."

THE END.

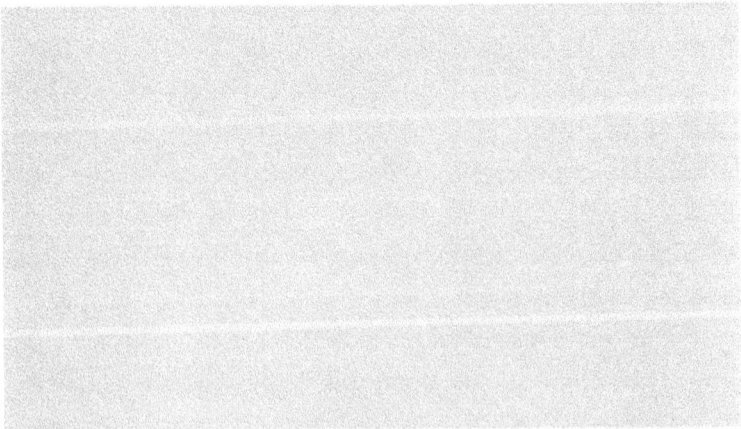

Cambridge : Stereotyped and Printed by Welch, Bigelow, & Co.

www.ingramcontent.com/pod-product-compliance
Lightning Source LLC
Chambersburg PA
CBHW021354210326
41599CB00011B/865